The Visible Kingdom of God

The Song of Noah

Esther Stein

BALBOA
PRESS
A DIVISION OF HAY HOUSE

Balboa Press books may be ordered through booksellers or by contacting:

Balboa Press
A Division of Hay House
1663 Liberty Drive
Bloomington, IN 47403
www.balboapress.com
1 (877) 407-4847

Print information available on the last page.

ISBN: 978-1-5043-9391-1 (sc)
ISBN: 978-1-5043-9392-8 (e)

Library of Congress Control Number: 2017919780

Balboa Press rev. date: 03/29/2018

Call to me and I will answer you and tell you great
and unsearchable things you do not know.
—Jeremiah 33:3

Soli Deo gloria.

Contents

Preface

When praying the Lord's Prayer, we ask for His kingdom to come. Have you ever wondered what the sky will look like when He answers that prayer? As the psalmist said,

> In the beginning you laid the foundations of the earth, and these are the work of your hands. They will perish, but you remain; they will all wear out like a garment. Like clothing you will change them and they will be discarded. (Psalm 102:25,26)

Our early ancestors tell us that the sky once looked very different. At the time of the Flood, the heavens changed like clothing. This is our common heritage.

Noah and his family told the story of a star, a creation of God unlike any other. Every tribe spoke of the perfect beauty of this star, which they pictured in the form of a cross. They called it the kingdom of their ancestral Father. They said it was seen in earth's sky continuously between the fall and the Flood. They said it never rose or set but was fixed directly overhead, the position of the present sun at noon. They called it the star of God because the Lord Himself illuminated it. They called it the lodestar because He held the earth in His loving embrace. It was the lone star because no other celestial objects could be seen in the light of His glory.

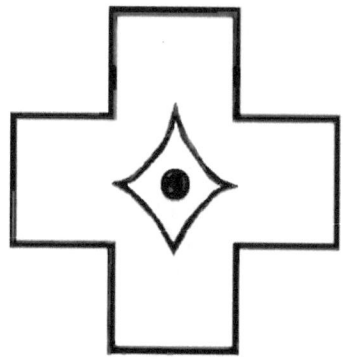

The Mayan "Great Star" glyph (ES)

Unlike the present sun, the light of the star of God was never blinding or burning but always healing and nurturing. The disappearance of His star was the greatest sorrow ever to befall humanity. It is no accident that the words cat*astro*phe and dis*aster* contain the word star. The root words *astro* and *aster* derive from the Greek *astron* (star). The loss of the Creator's visible presence caused unimaginable grief. All early prayers and rituals were designed to bring Him back. The star of God returned briefly on the first Christmas, where it shed its holy light over a small stable in Bethlehem. Look for the Lord's return and the reappearance of His star soon.

Chapter 1

The Sign of God

The Sign of the Son of Man

Then will appear the sign of the Son of Man in heaven. And then all
the peoples of the earth will mourn when they see the Son of Man
coming on the clouds of heaven, with power and great glory.
—Matthew 24:30

When Jesus was with his disciples on the Mount of Olives, they asked him
about the end of the age. He told them that just before His return to earth,
the sign of the Son of Man would appear in the sky and all nations of the
earth would mourn. What is "the sign of the Son of Man"? The first thing
that comes to mind is the cross. Has there ever been a sign more identified
with Jesus Christ than the cross? Did Jesus mean that a great cross would
appear in the sky announcing His second coming?

Celtic   Egyptian   Greek   Babylonian   American   Maltese   African (ES)

Prehistoric shrines show a deeply religious people who revered the cross
as their hope for eternal life. As the oldest and holiest of all symbols of the
then undivided human race, no other sign is so widely distributed in its most

basic form and with so many variants. The cross appears as the foremost sacred sign on rock art, cave walls, grave markers, and the earliest Neolithic pottery excavated. Whether worn as a protective amulet or woven into blankets and clothing, it adorned the robes of kings and commoners alike.

The cross is prominent in the design of ancient gardens, tombs, and sacred architecture. The oldest and largest prehistoric building in the world is the passage grave of New Grange, Ireland, built circa 3200 BC. Surrounded by a stone circle, the whole tomb chamber is in the form of a cross.[1] After the scattering of the people at Babel, the Creator had many different names, but His beloved sign remained. The cross is found in all early alphabets: Chinese, Egyptian, Etruscan, Linear A and B, Indo-Aryan, and proto-Sinaitic, one of the earliest Semitic scripts.

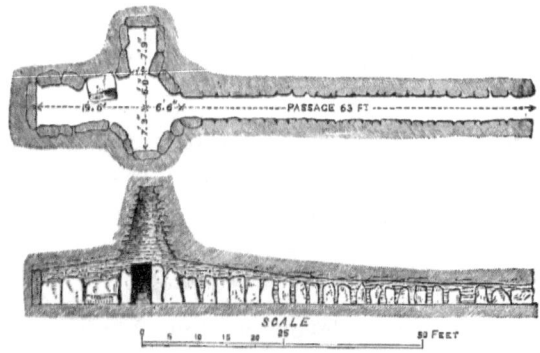

Plan and Section of Chamber in Newgrange Tumulus.

Interior of New Grange, County Meath, Ireland
After Wakeman's handbook of Irish antiquities (1903)

The Sign of God in Israel

Set up the tabernacle according to the plan shown
you on the mountain. —Exodus 26:30

The cross is fundamental to Israel and the Jewish people. It is the last letter of the Hebrew alphabet: *tav* (paleo-Hebrew +). According to the Jewish historian Josephus, the name *Hebrew* derives from Eber (or Heber), an ancestor of Abraham.[2] Eber in Babylonian is Neberu. In both languages, the name means "crossing place." Pictures of it appear on thousands of cylinder seals.

In tradition and ritual, the tav was *the* sign of God's protection associated with His throne of glory.[3] Jacob made the tav sign over his grandsons' heads (Genesis 10:24). The tabernacle was constructed in the form of a tav with three tribes in regiments at each of the four directions. All faced the tent of the congregation and the Ark of the Covenant in the center (Numbers 2:2). The camp of the Levites was laid out the same way.

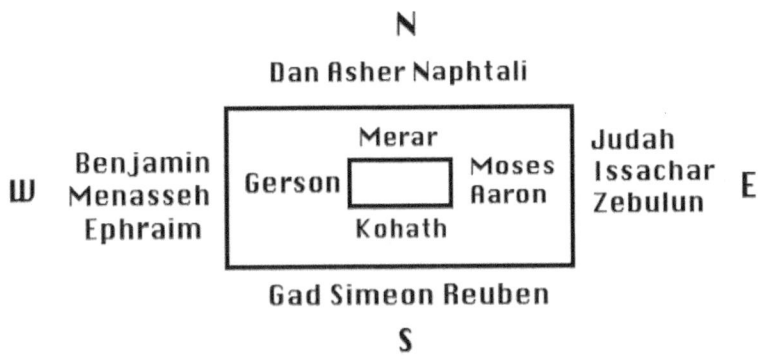

Artwork courtesy of Ava Raha.

The Lord commanded Moses to strictly adhere to the design shown to him on the mountain because it was "the heavenly pattern" (Exodus 25:40). The book of Hebrews reemphasizes this by saying, "They serve at a sanctuary that is a copy and shadow of what is in heaven. This is why Moses was warned when he was about to build the tabernacle, "See to it that you make everything according to the pattern shown you on the mountain" (Hebrews 8:5).

Biblical scholar Jack Finegan (1908–2000) wrote of the importance of the cross to early religion. The Talmud advises knowing the Torah from aleph to tav (Abodah Zarah 4a). The tav has the same relationship in Hebrew as the omega does in Greek. It is the sign of God and symbol of perfection. Written as + or **x**, the tav is the source of the Greek letters tau and chi, the first letter of Christos, and via Latin, the source of letters T and X. Job refers to the tav when he states, "Here is my mark" (Job 31:35). As the universal sign of God's protection, the tav was most probably the mark of Cain.[4] In rabbinic tradition, the sign God placed on Cain's brow was one of the twenty-two letters of the alphabet.[5]

In Ezekiel 9:4–6, the tav marks the foreheads of the faithful, the center of the forehead being the chosen site for ritual marks. From several passages in the Talmud, we learn that the priests of Israel were anointed with the tav or the Greek letter chi X.[6] The tav was also used by Jewish scribes to single out messianic passages in the Dead Sea Scrolls. At the beginning of the Christian era, Jewish believers knew the tav stood for "faithfulness, protection, and salvation."[7] The Hebrew dictionary defines tav + as "sign, mark, or cross," but after the adoption of the square Aramaic characters in the postexilic period, the tav no longer resembled a cross. It now looks like this: ת.

The mark on the forehead was also understood to be the tav in New Testament passages:[8] "Then I looked, and there before me was the Lamb, standing on Mount Zion and with him were one hundred forty-four thousand who had his name and his Father's name written on their foreheads" (Revelation 14:1). "They will see his face, and his name will be on their foreheads" (Revelation 22:4). Some important Hebrew words and phrases begin with tav:

> *tavnit* (pattern, copy, form, plan, replica)
> *tavnit HaMishkan* (pattern of the tabernacle)
> *Tavo Malchutechah!* (Thy kingdom come!)

The Morning Star

The Star of God on prehistoric Egyptian pottery, W M Flinders Petrie (1921)

> Just as I received authority from my Father, I will also give him the morning star. He who has an ear let him hear what the Spirit says to the churches.
> —Revelation 2:27b–29 (NKJV)

Although the cross is the most likely symbol for the sign of the Son of Man, another symbol is closely identified with Christ's appearing: a star. The magi recognized a particular star as the sign that a great king had been born in Bethlehem. They asked Herod, "Where is the child who has been born king of the Jews? For we observed his star at its rising, and have come to pay him homage" (Matthew 2:2). St. Ignatius of Antioch (ca. AD 35–107), one of the early church fathers, described it:

> A star shone forth in heaven above all the other stars, the light of which was inexpressible, while its novelty struck men with astonishment. And all the rest of the stars, with the sun and moon, formed a chorus to this star, and its light was exceedingly great above them all. And there was agitation felt as to whence this new spectacle came, so unlike to everything else [in the heavens].⁹

Since the star of God appeared at His Son's birth, shouldn't we expect its reappearance at His second coming? In Revelation, Jesus promised to give the morning star and authority over the nations to those who overcome and do His will to the end. In the last chapter, He identifies Himself with the morning star. "It is I, Jesus, who sent my angel to you with this testimony for the churches. I am the root and the offspring of David, and the bright morning star" (Revelation 22:16).

In his Olivet discourse, Jesus warned that the nations will be in great distress at the end of the age. There will be signs and wonders in the heavens (Luke 21:11, 25). Will the greatest of these be the reappearance of His star announcing His imminent arrival? The observance of a bright new star in the heavens would certainly cause a sensation, but it might not be perceived as a sign of the return of Christ as much as an unusual astronomical event. Yet if the two signs are combined, it would definitely alarm the nations.

After a comprehensive study of ancient religion, Mesopotamian scholar Stephen Langdon asserts that the star cross is "the *only* religious symbol of the primitive period." It means "god, star and heaven" and is found on cylinder seals all over Mesopotamia!¹⁰ This applies to the entire Neolithic period in general in every part of the world. The ancient religious literature of India speaks of this star,

> Who urged the high and mighty sky to motion, the
> Star of old, and spread the earth before him.
> —*Rg Veda* book 7, Hymn 86
> Ralph TH Griffith, trans. *Hymns of the Rg Veda* (London 1889)

The Seal-Cylinders of Western Asia. William Hayes Ward. (1835–1916)

The Egyptian word for God, *neter* ⌐|, is a temple flag. Neter also means "morning star." Neter Ta is "the land of God." Neter is the origin of the Latin word *natura*, from which the word *nature* is derived.[11] Our ancient ancestors speak with one voice of the awesome glory of this star, whose fourfold rays illuminated the entire antediluvian world. They saw it as God's dwelling place and the desired destination of every soul. They continually mourned His disappearance at the Flood and prayed earnestly for His return.

The holiness of this star is evident in the word itself. In Hebrew, the name of God—*El*—occurs in Helel twice. Helel הֵילֵל (morning star) and *halel* הֵילֵל (praise) are identical in the original Hebrew, because when the people looked up and saw His beautiful star, they praised God. "I lift up my hands toward your Most Holy Place" (Psalm 28:2).

Similarly, the Egyptian word *tua* (morning) also means "to praise, adore, honor."[12] Its determinative sign (the earliest, most instructive clue to a word's meaning) is a person lifting up hands in praise 𓀃. Tua ur ⌐𓃾𓏏★𓆳𓀃 is the "great star of the morning." Tua Neter ⌐|⌐𓃾 is "the star of the God." ⌐|★𓋹 is "to thank God." Tua-t ⌐𓃾𓈖 is "the Otherworld," called "everlasting" and "hidden."[13] The Pyramid Texts, the world's oldest hymns, speak of the morning star but *not* the evening star. The word *tua* (morning) includes Ua (the One God) 𓈖𓂀𓃾. The first glyph is the original form, the second is the late form. Ua also means "to drive away" and "going to ruin."[14]

The ancient Persians revered a special star, the bright and glorious Tishtrya, the first star, the lord of all stars. More than a celestial object to them, Tishtrya was their beneficent protector and provider who presided over time. He produced the waters at the beginning of the creation and was the source of all moisture and fertility. Each drop of water he produced was as big as a bowl so that the earth was covered with water. Tishtrya also rained down seeds over the earth.[15]

The Aztecs claimed the Toltecs as their cultural forebears. Ce Acatl, the morning star, was their hero first ruler of the sacred homeland Tula.[16] He was a preexistent power older than the sun that cast forth glittering rays. He was dawn itself. He fell at the time of the great Flood, a cataclysm that brought ruin to the people. This was their greatest sorrow, when the star of stars sank down into the earth. They said this doomed star rose again in splendor but disappeared for good after losing a duel with the new sun.[17]

All over the world, explanations arose to account for the disappearance of the star of God. A folktale from the Brazilian highlands blames an unfaithful spouse. The beautiful, brilliant Star Woman asked her husband to plant a garden where she could sow many nourishing crops from the sky. She sent down yams, corn, potatoes, rice, and peanuts. But Star Woman's husband was untrue to her. Greatly offended, Star Woman rose into the sky and never returned. If her husband had been faithful, the people would still have all the wonderful things of heaven.[18]

While lamenting the loss of the star of God and praying for His return, our ancient ancestors annually performed memorial rituals. According to the Fiote people of Africa's Loango Coast:

> The Star Way is the road for a funeral procession of a huge star,
> which, once, shone brighter from the sky than the sun.
> —E. Pechuel-Loesche, Volkskunde Von Loango (1907)

Pharaoh with arms crossed over the heart
Artwork courtesy of Ava Raha

The Sign of God in Egypt

The Lord Almighty will bless them, saying, "Blessed be Egypt my
people, Assyria my handiwork, and Israel my inheritance."
—Isaiah 19:25

The ancient Egyptians, descendants of Noah through Ham, had many
variations of the cross. The arms of deceased pharaohs always cross over the
heart. Un ⚒ (to be, to exist) is a picture of the One God hovering over the
primeval waters. ⚒ means "I Am God."[19] The Ankh cross ☥ stands for "life,
stability, joy of heart." Ankh-t is a name for heaven, and the Ankhu are "the
beatified in heaven."[20] The star of Ari ⚛ has twelve rays, three at each of the
cardinal points, the same arrangement as the camp of Israel.[21]

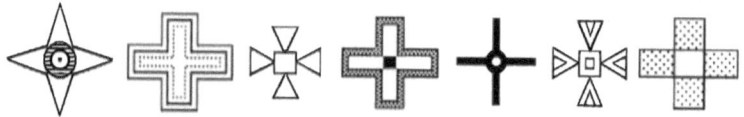

North West Coast, Alaska, and Arctic regions (ES)

The Sign of God in America

When He prepared the heavens, I was there: when
He set a compass upon the face of the deep.
—Proverbs 8:27 (KJV)

Native Americans revere the morning star and the cross as symbols of their
Great Spirit. The Arapaho say the star-child gave his image to the world as
a cross and became the morning star.[22] The cross is the Apache and Piman
symbol of their great sun-father. The Wintun of California tell of a new sky
that replaced the old one supported by pillars at the four cardinal points.
The Muskogean (Creek) build four lodges open to the central square for
their New Year rites. Each of the four lodges has three compartments facing
a central fire.[23]

The Pawnee say the star of Tirawa Atius founded their villages. He is the Supreme Father Heaven whose abode is the highest circle of the visible universe. Tirawa Atius is His Pawnee name. All the powers in heaven are derived from him; He is the father of all things visible and invisible, and father of all the people.[24]

Their chief is his representative, and their earth lodge is aligned to the four directions. Its posts represent the morning star and North Star. Like ancient Israel, the Pawnee arranged their camp to reflect his mystic sign. The signs x and + are both symbols for star.[25]

Tirawa held council with the other gods and gave each of them assignments. Death was introduced by a lesser star, jealous of all the favors Tirawa bestowed on a very bright star, which played a large part in the creation. This jealous star found a "sack of storms" that Tirawa had entrusted to the bright star. He then emptied its contents and sent terrible storms to the earth. This started a raging fire until Tirawa put it out with a catastrophic deluge. A man and his wife survived with some maize, pumpkin seeds, a drum, and a pipe. They restarted civilization.[26]

When the Mayan city of Palenque was excavated in 1952, archaeologist Alberto Ruiz Llullier discovered a passageway under the floor of the main chamber of the Temple of the Inscriptions. After three seasons of clearing away rubble, they found a one-hundred-foot stairway leading to a wall twelve feet thick. After a week of backbreaking labor, they broke through, uncovering funerary offerings, including the bodies of several young men. Beyond the blocked north wall lay a sight unseen in a thousand years: a twenty-ton sarcophagus. On the magnificently carved lid was the figure of the ruler Pakal the Great.[27] He was looking up at the cross, his hope of eternal life. Bordering the slab are glyphs of the morning star and the sky-throne of his God.[28]

The cross was the sign of God in South America long before Christianity arrived. In a creation account of the Apapocuva Guarani of Brazil, the fate of the earth depends on it:

> Our Great Father came alone, in the midst of the darkness he disclosed his presence alone. The eternal bats fought with one another in the midst of the darkness. Our Great Father had the sun in his breast. And he brought the eternal wooden cross. He laid it in the direction of the east, trod upon it, and began to make the earth. To this day the eternal wooden cross remains as the earth's support. As soon as he removes the earth's support, the earth will fall. Then he brought the water.[29]

William Hayes Ward, *The Seal Cylinders of Western Asia*
Carnegie Institution of Washington (Washington, DC, 1910)

The Crossroads of Life

You should not wait at the crossroads to cut down their fugitives,
nor hand over their survivors in the day of their trouble.
—Obadiah 1:14

The four arms of the cross are described in many picturesque ways.[30] Often they are just "the crossroads." The Mayan creation text, Popol Vuh, speaks of *Cahib xalcat be* (four junction roads) each shining with a different color: red, black, white, and yellow.[31] Temples and shrines were built at crossroads. As places of divine justice, criminals were tried and executed there. Divination was performed at crossroads, "For the king of Babylon will stop at the fork in the road, at the junction of the two roads, to seek an omen" (Ezekiel 21:21a).

Cicero described the Romans sacrificing annually at the *compiti* (from *compitum*, meaning "crossroads") during their ancient feast of the Compitalia in January. The guardians of crossroads were the Lares (ancestral spirits).[32] Similarly, the Chimata-no-Kami are deities of crossroads in Japan. Offerings are placed, and ceremonies are conducted there twice a year.[33] The ancient festival Michi-ae-no-matsuri (Festival of the Road Gods) is celebrated at the four corners of the capital. Protection is sought from angry ancestral spirits and demons.[34]

The Four Winds

I have scattered you to the four winds of heaven, declares the Lord.
—Zechariah 2:6b

When Jesus was describing the events of the last days prior to His return to earth, He spoke of the four winds of heaven. "They will gather his elect from the four winds, from one end of heaven to the other" (Matthew 24:31). This would have been readily understood by our ancient ancestors, who saw them as the breath of the Great Spirit blowing across the face of the deep (Genesis 7:11, Proverbs 8:22). The prophet Daniel also mentions them. "In my vision at night I looked, and there before me were the four winds of heaven churning up the great sea" (Daniel 7:2).

The ancient Greeks sacrificed to the four winds.[35] When a new pharaoh assumed the throne of Egypt, he took up his bow and shot arrows to the four directions.[36] This is a prime example of how cosmic imagery assimilated into earthly imagery, how succeeding generations replaced what *was* seen for what *is* seen. This ritual lives on in the present. Prince Akihito of Japan performed this rite as part of his coronation as emperor in 1989. It is called Shi-Ho-Hai (Four Directions Worship).[37]

The Dayspring on High

Through the tender mercy of our God with which
the Dayspring from on high hath visited us.
—Luke 1:78 (NKJV)

Genesis 2:10 says, "A river watering the garden flowed from Eden; from there it was separated into four headwaters." This dayspring on high is revered in all ancient cultures. In India, it is known as Svarnara (bright spring, lord of heaven). In China, it is 源泉 *yuánkuán* (fountainhead). The Norse heaven, Asgard, had four milk-white streams flowing from Audhumla (void).[38] In the old runic alphabet, x *gyfu* (gift) also means "sacrifice."[39]

The Egyptian hieroglyph for city or town is . It is a picture of the crossroads of life, but it is also known as the Four Niles. It is one of the most significant pictorial glyphs in the Egyptian language. Egyptologist Heinrich Brugsch recognized the Four Rivers of Paradise in this glyph,[40] and author

William F. Warren called it "a pictorial symbol of the primitive Eden divided by its fourfold river."[41]

Ea with the Four Rivers flowing from him
The Adda Seal, Akkadian cylinder seal (circa 2300 BC)

The cities of ancient Mesopotamia all claimed their god was the Creator. According to British Assyriologist Stephen Langdon, "All Semitic tribes appear to have started with a single tribal deity whom they regard as the divine creator of his people, and this deity seems to have been astral."[42] Ea was the God of Eridu, known as Enki in Sumerian. The name is a cognate of the Hebrew Yah and Akkadian Ay(y)a (to live).[43] Ea created all things by his word, and his paradise, Dilmun, is the garden at the mouth of the rivers. Ea warned Utnapishtim and his family to prepare for the Flood. With his help, they survived with the seed of all living creatures. Utnapishtim was eventually taken to live forever at the mouth of the rivers.[44] This is the crossing point where the Lord of heaven dwells.

Mayan

Roman (ES)

In some traditions, rather than four rivers flowing from a central point as in Genesis, there are two heavenly streams, one crossing over the other. The Babylonians called their two local rivers after the celestial Tigris and Euphrates. The Chinese originally knew two celestial rivers with no beginning or end: Tiānhé 天河 and Tiānhàn 天汉[45] (although now both are interpreted as the Milky Way). The ancient Persians also recognized two heavenly rivers, the Arag and the Veh.[46]

In India, Brahman, the supreme spirit, is the Creator of the universe. His day and night is one thousand years.[47] He dwells in the confluence of two heavenly rivers of grace: the white River Ganga (of salvation) and the black River Yamuna (of origins). These rivers are "holy and sin-destroying" and appear on the doorjambs of ancient temples.[48] Although it is stated in the hymns of the Vedas, "Of him whose glory is so great, there is no image," Brahma has been depicted cross-legged with four faces and four arms. Four is the number of totality and perfection in India.[49]

Plato's description of God forming the cosmos is now clear:

> This entire compound he divided lengthways into two parts, which he joined to one another at the center like the letter X, and bent them into a circular form, connecting them with themselves and each other at the point opposite to their original meeting-point. ("Timaeus," trans. Benjamin Jowett 1817–1893)

Similar statements are made in the sacred texts of Mesopotamia:

> Once on a time, when Ninurta decreed fates then in the Land lived the X stone, it is said. Verily this is so.
> —Stephen Langdon, *Semitic Mythology*
> *Mythology of All Races*, Vol. 5 (1916)

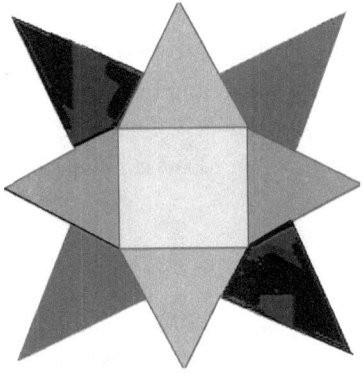

Star on the marble floor of the Second Temple, Jerusalem (ES)

Two rivers crossing over one another are also found in the Americas. The Navajo call the blue world "the Place of the Crossing Waters," where a cool north-south stream passed under a warm east-west stream.[50] The Pawnee know it as "the Place of the Division of the Waters."[51] In the cosmology of the Sia of New Mexico, the Creator Sussistinnako drew a cross of meal where he dwelt, a north-south line crossed by an east-west line. On each side of the line, he placed two precious parcels. Then he sat down and began singing low, sweet music. People, animals, and birds began to appear, and he continued to sing until his creation was finished.[52]

The Four Pillars of Heaven

The pillars of the heavens quake, aghast at his rebuke.
—Job 26:11

In the Bible, the four rivers are also called "the pillars of the heavens." The four golden pillars supporting the curtain separating the Holy Place from the Holy of Holies represent them. "You shall hang it on four pillars of acacia overlaid with gold, which have hooks of gold and rest on four bases of silver" (Exodus 26:32). Josephus explained that within the four pillars is a heaven peculiar to God:

> Now the whole Temple was called the Holy Place, but that part within the four pillars, and to which none were admitted, was called the Holy of Holies.
> —Josephus, *Antiquities of the Jews* book 3, chapter 6

The psalmist calls them "the beams of his chambers in the waters" (Psalm 104:3). The book of Enoch mentions them: "And I beheld the winds occupying the height of heaven arising in the midst of heaven and of earth, and constituting the pillars of heaven" (Enoch 18:4, 5).

The Egyptians originally placed these pillars at the compass points:

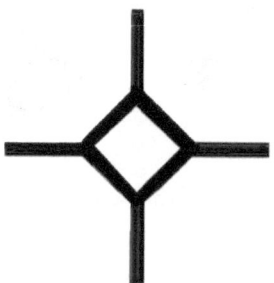

> The Egyptians first divided the world into four parts, each corresponding to one of the four pillars, which held up the sky, that is to say, to one of the four cardinal points, East, South, West, and North. (EAW Budge, *The Egyptian Book of the Dead*, preface)

The Egyptian word *skhenut* (pillar) is very similar to the Greek word *skenoo* σκηνόω (to pitch a tent, to tabernacle, to dwell) used in "He who sits on the throne will spread His tabernacle over them" (Revelation 7:15). Later, the scribes represented the *skhenut* with the sky slipping down over them ⊓⊓⊓.[53] These pillars of heaven are imitated everywhere in ancient architecture. They appear on the façade of early temples, two on each side of the entrance.

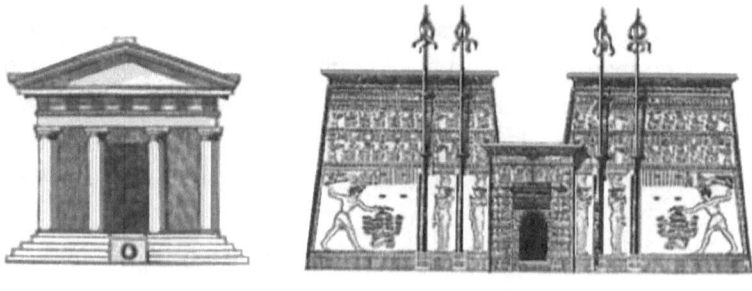

Greek   Egyptian

Encyclopedia Britannica (1911)

In the Mesopotamian texts, the pillars are the crossbars of a roof, or the canals the gods created as lifelines of the land. We often find them mentioned by the classical authors:

> Who knows the depths of every sea, and himself holds
> the tall pillars which keep earth and heaven apart.
> —Homer (eighth century BC), *Odyssey,* trans. AT Murray (1919)

The oldest hymns of India, the Four Vedas, refer to them:

> Who with great wisdom measured both the regions out, and
> established them with pillars that shall ne'er decay.
> — *Rg Veda* book 1, Hymn 160,
> trans. Ralph TH Griffith, *Hymns of the Rg Veda* (London 1889).

They are also found in national epics, such as the Finnish *Kalevala*.

> Rocks were fastened in the ocean;
> Pillars of the sky were planted.
> —Rune 1 trans. John Martin Crawford (1888)

The Sign of God in East Asia

> For the son of man in his days will be like the lightning, which
> flashes and lights up the sky from one end to the other. But first
> he must suffer many things and be rejected by this generation.
> —Luke 17:24, 25

The Japanese Kojiki (Record of Ancient Matters) calls them "the crossbeams of Takama-No-Para" (Plain of High Heaven). Representations of them—Onbashira-sai—are the four sacred poles raised at the four corners of the Japanese Shinto shrine grounds every seven years. They are taken from the sacred forest at Suwa Taisha.[54] In China, it is said that the circle of four pillars, or props known as Sse Fu, retained the celestial waters until its breach caused the Flood.[55] *Fu* 父 means "Father" in Chinese, and Tiānfù 天父 is our heavenly Father. Fu means "four" in Egyptian.

The cross is the ancient Chinese character *shi* 十 symbolizing the four cardinal points and center.[56] It represents the first ancestor whose symbol, a cross, appears on early bronzes.[57] As a word, *shi* means "perfect" and the number ten like the Latin X. According to the sage Liu Sheh, "The entire universe is included in this sign; the origin, existence and infinite renewal of life."[58] *Shi* under *Fu* 父 between the ancient characters for heaven and earth forms *sheng* 圣 (holy). The ancient character *wáng* (king) 王 is the sign of God between heaven and earth, as he mediates between them.[59]

The Star of God on Chinese Neolithic pottery, 6000–1000 BC (ES)

The Ostiak Samoyeds describe "a dwelling place of the gods … on four high supports." The Siberian Kalmuks remember four rivers with jewel sands of different colors. The eastern has silver, the southern has blue jewel sand, the western has red jewel sand, and the northern has gold sand. Among the Altai Tatars, the appearance of a new star signifies birth and the falling of a star represents death.[60]

The Sign of God in the South Pacific

Behold, He is coming with clouds, and every eye will see
Him, even they who pierced Him. And all the tribes of the
earth will mourn because of Him. Even so, amen.
—Revelation 1:7

The original inhabitants of Hawaii worshipped Teave (tay-yah-vay), who brought order out of chaos by breathing forth mana (vital spirit energy). His holy sign, the cross, represents divine grace and resurrection.[62] One of his sacred titles was I'ao (Eh-yah-ho) meaning, "Infinite World, Infinite Light," as he was the supreme light of the world. He is called Io in other parts of Polynesia. Tau is an ancient South Sea Island word for the scepter of God, a high cross. Every king and high priest had his own Tau to carry in religious processions and temple rituals.[63] A great white bird called Halulu represents the Polynesian Supreme Spirit.

Though names may differ among the Australian aboriginal tribes, symbols and beliefs are similar throughout the continent. In ceremonies and song cycles, the morning star is a symbol of eternity and represents life after death.[64] All tribes believe in a benevolent supreme being who existed before death. They call Him All-Father, Master, or Maker. The dreamtime or "dreaming" is the traditional name for the sacred era that began with a void. According to renowned historian of religion Mircea Eliade, these beliefs are archaic and indigenous and long preceded the arrival of any missionaries.[65]

The Kurnai of southeast Australia believe in Mungan-ngaua (Father of Us All) who has a voice resembling thunder. They say he is everlasting, as he existed from the beginning, and he still lives. During his initiation, every young man learns the sacred drama of his creative deeds, his gifts and laws, his anger and disappearance. The dreamtime ended when He brought on a cosmic cataclysm in which almost the entire human race perished. The present fallen world is the result of the sin of a primeval ancestress. The people received instructions on how to live from "the authors of the disaster."[66]

Jesus Christ, the First and Last

This is what the Lord says—Israel's King and Redeemer, the Lord Almighty: I am the first and I am the last; apart from me there is no God.
—Isaiah 44:6

Paleo-Hebrew   Modern Hebrew

The New Testament is written in Greek, and most Bibles use the Greek letters alpha and omega to translate Revelation 22:13. Jesus did not speak Greek to John. He spoke Aramaic, a Hebrew dialect. He said, "I am the Aleph and the Tav, the beginning and the end, the first and the last" (Revelation 22:13 ONMB). The Aleph-Tav appears thousands of times in the Tanakh (Hebrew Bible), but the translations we commonly use leave it out. Rabbinic scholars call it the word of creation. It is pronounced εt.[67]

The Aleph-Tav is in the very center of Genesis 1:1, like the servant candle of the menorah, the one that lights all the others and never goes out. Here is Genesis 1:1 in the Tanakh. The original Hebrew text forms a menorah with three words on each side of the Aleph-Tav.

הַשָּׁמַיִם וְאֵח הָאָרֶיִים אֵח הַשָּׁמַיִם וְאֵח הָאָרֶץ

In the beginning created God Aleph-Tav the heavens and the earth.

The Aleph-Tav אֵח is centered between Yeshua (Jesus) and Moshiach (Messiah) in "You went forth to save your people, to save (Yeshua) אֵח anointed one (Moshiach)" (Habakkuk 3:13a). "They have forsaken the אֵח Lord, they have provoked the אֵח Holy one of Israel to anger" (Isaiah 1:4). "And I will pour on the house of David and on the inhabitants of Jerusalem the Spirit of grace and supplication; then they will look on Me אֵח whom they pierced. Yes, they will mourn for Him as one mourns for his only son, and grieve for Him as one grieves for a firstborn" (Zechariah 12:10).

Every tribe once expected the reappearance of the morning star at the end of the age. The Arapaho believe their ancestral Father will wear the morning star when He leads the resurrection of their ancestors.[68] After the Flood, Venus, the present morning star, was singled out for special observation. Trained astronomer-priests tracked its movements. Although it bears no resemblance to the star of God, whole societies organized their lives around its appearance and disappearance. Since the star of God belongs to the entire human race, when it appears again, every nation will recognize it.

The Yucatec Mayan book of Chilam Balam prophecies:
At the end of the thirteenth age, the signal of God will
appear on the heights, and the Cross with which the
world was enlightened, will be manifested.
— Lewis Spence, *Arcane Secrets & Ancient Lore of
Mexico & Mayan Central America* 256.

In early Christian art, Jesus wears His kingdom behind His head,

Basilica of Sant'Apollinare Nuovo
Ravenna, Italy

Chapter 2

The Throne of God

Heaven Is My Throne

And he who swears by heaven, swears by the Throne
of God and by Him who sits on it.
—Matthew 23:22 (NKJV)

When Jesus called Heaven "the Throne of God," He was quoting the prophet Isaiah. "This is what the Lord says: 'heaven is my throne and the earth is my footstool. Where is the house you will build for me? Where will my resting place be? Has not my hand made all these things, and so they came into being?' declares the Lord" (Isaiah 66:1, 2a). Since Heaven is God's throne, we must view the terms interchangeably. Whatever is said about one is true of the other. A throne is a high stationary seat where a king sits, often made of gold and studded with gems. This is a good description of God's Throne, the prototype of all royal thrones.

We know from their testimony that the Throne of God was engraved on the minds of the Flood survivors. Jeremiah wrote, "A glorious throne, exalted from the beginning, is the place of our sanctuary" (Jeremiah 17:12). They only had to look up and see His throne shining forth in awesome majesty. They literally walked in the light of His presence (Psalm 89:14–15b). There was communication between Heaven and earth undreamed of today. No sun, moon, or stars were seen, or even needed, because God supplied all necessary light and warmth. Everything required for life and happiness was abundantly provided.

The Rider on the Ancient Heavens

Sing to God, you kingdoms of the earth; Oh, sing praises to the
Lord, to Him who rides on the heaven of heavens, which were
of old! Indeed, He sends out His voice, a mighty voice.
—Psalm 68:32, 33 (NKJV)

The Ancient of Days holding a dove riding on the ancient heaven.
Coin minted in Judea, fourth century BC.
—Stephen Langdon, *Semitic Mythology*
Mythology of All Races, Vol. 5 (1916)

On the Judean coin above, the Creator sits on a winged wheel, which our
ancient ancestors called the primal wheel. It appears on the insignias of many
nations and has many titles: the celestial wheel, the potter's wheel, the wheel
of existence, the wheel of law, the wheel of virtue, the indestructible wheel of
the cosmos, the wheel of truth, the wheel of righteousness, the wheel of birth
(James 3:6b AMP). The Bible calls it the threshing wheel (Proverbs 20:26).
In ancient symbolism, the central space, where the radii meet at its axis, is
where God resides.[1]

The wheel, one of the oldest symbols of God, was seen as His own throne-
chariot (Hebrew Merkavah, Aramaic Merkabah). The word is related to the
Egyptian *khabs* (star) and *khab* (chariot). This is why David calls Him "the
rider on the ancient heavens" (Psalm 68:33). Moses also refers to Him in this
way, "There is no one like the God of Jeshurun, who rides on the heavens
to help you and on the clouds in his majesty" (Deuteronomy 33:26 KJV).
Similar phrasing is found in the Dead Sea Scrolls:

The cherubim bless the image of the Throne-Chariot above the firmament, and they praise the majesty of the fiery firmament beneath the seat of his glory. And between the turning wheels, angels of holiness come and go, as it were a fiery vision of most holy spirits; and about them flow seeming rivulets of fire, like gleaming bronze, a radiance of many gorgeous colors, of marvelous pigments magnificently mingled.
—*The Complete Dead Sea Scrolls in English* (trans. Géza Vermès)

The word *universe* literally means "the One that turns." The throne-chariot was the entire universe to the antediluvians as seen in the following ancient symbols:

| Phoenician | Mayan | Norse | Chinese (ES) |

All early tribes associated their Creator with the wheel. In the Upanishads of India, it is said of the One Supreme Spirit:

Some wise men speak of Nature, and others of Time as the cause of everything, but it is the greatness of God by which this Brahma-wheel is made to turn.
—Svetasvatara Upanishad (trans. Max Müller)

The *Rg Veda* calls the righteous Varuna "the nave set within the wheel in whom all wisdom centers" (Book 8, Hymn 41). Vishnu's wheel had a nave of thunder. The Mahabharata calls Indra a wheel-turner king:

And the glorious Wheel of the great spirited Bharata rolled thundering through the worlds, grand, radiant, divine, unvanquished … He was a king, a Turner of the Wheel, a majestic worldwide monarch. He sacrificed many sacrifices, he was an Indra, Lord of the winds.
—The *Mahabharata* (trans. JAB van Buitenen)

Savitar (generator, nourisher), the Creator-God of Bihar and Tamilnad, is the impeller-God of beginnings ❀, the driving force of the universe.[3] The word *impeller* derives from the Indo-Aryan root *pel*, meaning "to drive."

The Wheel King

As I looked, thrones were set in place, and the Ancient of Days took his seat. His clothing was as white as snow; the hair of his head was white like wool. His throne was flaming with fire, and its wheels were all ablaze.
—Daniel 7:9

The ancient sacred texts of the Far East describe the reign of the Chakravartin (the wheel-turning emperor). He once ruled the earth, not with the sword but by the power of righteousness:

> Long, long ago, brethren, there was a sovran (sic) overlord named Strongtyre, a righteous king ruling in righteousness, lord of the four quarters of the earth, conqueror, the protector of his people, possessor of the seven precious things, to wit: the Wheel, the Elephant, the Horse, the Gem, the Woman, the House-father, the Counselor. More than a thousand sons also were his; heroes, vigorous of frame, crushers of the hosts of the enemy. He lived in supremacy over this earth to its ocean bounds, having conquered it, not by the scourge, not by the sword, but by righteousness ... But on the seventh day after the royal hermit had gone forth, the Celestial Wheel disappeared. (*Cakkavati Sihanada Suttanta*, trans. TW Rhys Davids 1881), *Sacred Books of the East*, edited by Max Müller,

The Lord moves His throne-chariot wherever and whenever He wishes and makes it appear and disappear at will. This is one of the reasons why our ancestors considered Him a God of magic and mystery. Ezekiel saw His throne-chariot coming out of a whirlwind in the north and described it as resembling gleaming metal and sapphire (Ezekiel 1:26–28). His formerly visible throne-chariot may still be in His chosen place but hidden from our view (i.e., Egyptian *Amen-t*, Italic *Latium*, ancient names for Heaven meaning "hidden, concealed").

The following reflects the ancient beliefs of Malaysia:

> The Creator of the entire universe preexisted by Himself, and
> He was the Eldest Magician. He created the earth of the width of a tray
> and the heavens of the width of an umbrella, which are the universe

of the Magician. Now from before the beginning of time existed that Magician that is God and He made Himself manifest with the brightness of the moon and the sun, which is the token of the True Magician.
—Walter W. Skeat: *Malay Magic: An introduction to the Folklore and Popular Religion of the Malay Peninsula* (1900)

The ancient Chinese sages regarded the chariot wheel as the symbol of authority and sovereign rule.[4] They assert that the soul climbs onto the wheel of life and death. Flames of five colors sprang from it, and its glow lit up the heavens.[5] The Greek authors spoke of wheels suspended in Egyptian temples, and there was similar symbolism in the medieval chapels of France. Wheels were revolving altars hung from the roof or set high on pillars. With bells attached, cords pulled by the worshipper set them turning for "the edification of the faithful."[6]

The Wheel of Destiny

L'Hortus Delicarum. Herrad of Landsberg, twelfth century. ⚠

The impact of this archaic symbol on Noah's descendants cannot be adequately appreciated today. The wheel[1] was once considered so holy it was not put to practical use until much later when the terror of the Flood had subsided, lest it offend the awesome driver who made the sky fall. Numerous examples of

---

[1] There seems to be an archaic linguistic relationship between the Sanskrit *chakra* (wheel) and *sakara* (brilliant rays), the Latin *sacra* (holy), the Inca *chacra* (dwelling house), the Hebrew *shakan* (to dwell), and *shachar* (morning star, dawn).

ancient customs reenacting the loss of the king and his wheel kingdom exist far and wide, including this one from Germany as recounted by Frazer:

> At Eisenach on the fourth Sunday in Lent young people used to fasten a straw-man, representing Death, to a wheel, which they trundled to the top of a hill. Then setting fire to the figure they allowed it and the wheel to roll down the slope. (James G. Frazer, *The Golden Bough*, chapter 28 ABR)

The festivals of the earliest nations featured chariot wheels decorated with flowers. The use of spinning wheels was prohibited at certain times, especially at the winter solstice.[7] Memories of the primal wheel survive in the fairy tales of *Sleeping Beauty* and *Rumpelstiltskin*. Every king once wore a "king's wheel" with precious stones at the hub and four cardinal points. These have been found in Bronze-Age tombs in Ireland, some dating to 1500 BC.[8]

On the subject of the loss of the primal wheel, Shakespeare laments the catastrophic collapse of the whole world order in Rosenkrantz's speech "The Cease of Majesty."

> It is a massy wheel fix'd on the summit of the highest mount,
> To whose huge spokes ten thousand lesser things
> Are mortis'd and adjoin'd; which when it falls,
> Each small annexment, petty consequence
> Attends the boist'rous ruin.
> Never alone did the King sigh, but with a general groan.
> (*Hamlet,* act III, scene 3)

Jesus said that His kingdom "was not of this world" (John 18:36) and spoke of traveling to "a far country" (Mark 12:1, 13:34). The Kingdom of God may have been out of sight, but it was never out of mind. The Egyptians hoped their rituals would bring back the Sekh't-aanru, the beautiful region of heaven where "the souls of the blessed dead lived and served Osiris).["9] As we are all descendants of Noah, the primal wheel was the symbol for the land of God in every tribe.

It is the original source of certain letters appearing in Old Hebrew and Phoenician as the letter *tet* ⊕. This sign, known as "the compass," is part of all early alphabets, including Mayan, Chinese, Linear A & B, Etruscan, and the Indus Valley scripts. It also appears on early rock paintings all over the world. The *tet* became the Greek *theta,* originally written $\oplus$ (now θ). Theta is the first letter of God—Theos—and His Throne—Thronos.

Astonishing Similarities in Ancient Descriptions of Paradise

> With mighty chariotry, twice ten thousand, thousands upon
> thousands, the Lord came from Sinai into the holy place.
> —Psalm 68:17

Descriptions of the land of God, whether oral, written, or drawn, are identical in almost every detail, because they all stem from a common source, the One, who the Greek tragedian Aeschylus (c. 525–456 BC) described as "the Divinity that holds his sacred throne in strength, above the sky."[10] The ancient sky magnified the Lord, and every tribe remembers that glory. Before we examine their amazing descriptions, let's review some biblical facts about Heaven, God's throne.

> Heaven is in the north (Job 37:22; Isaiah 14:13; Psalm 48:2).
> Heaven is encircled with a rainbow of seven colors (Ezekiel 1:28; Revelation 4:3).
> Heaven is set on a high mountain (Psalm 48:1, 2; Isaiah 2:2, 3; Hebrews 12:22).
> Heaven is like a king who gave his son a wedding banquet (Matthew 22:2–3).
> Heaven is for those written in the Lamb's Book of Life (Revelation 13:7–8, 21:27).
> Heaven is a land that is (now) very far off (Isaiah 33:17, Mark 12:1, 13:34).
> Heaven has come in the person of Christ (Matthew 4:17; 10:7).
> Heaven is for those who are born again (John 3:3, 3:7, 1 Peter 1:23).
> The Lamb is at the center of the throne (Revelation 7:17).

The capital city of Heaven is the New Jerusalem.

> The city is 12,000 stadia in length (1,500 miles). It is seen from earth as a square, but it is actually a perfect cube (Revelation 21:15, 16).

God and the Lamb are the temple within it (Revelation 21:22).
The city is pure gold. Its jasper walls are 144 cubits thick (Revelation 21:17–18).
The city has the glory of God and the Lamb for its light (Revelation 21:23).
The wall foundations are all kinds of precious stones (Revelation 21:19–20).
The city has a fountain of living water (Isaiah 49:10; Revelation 7:17; 21:6).
The city has a healing tree of life in the center (Revelation 2:7, 22:2).

Israel

In love a throne will be established; in faithfulness a man will
sit on it—one from the house of David—one who in judging
seeks justice and speeds the cause of righteousness.
—Isaiah 16:5

The early rabbis described Jerusalem as gem studded,

Seven walls will encompass Jerusalem, which will be composed of silver, gold, jewels, bright stones, sapphire, carbuncle, and fire; and its brilliance will cast light from one end of the world to the other. The Temple will be built on four mountains (from) refined gold, purified gold, beaten gold, and the gold of Parvaim, like gold which produces fruit, set in sapphire (and) fixed in bdellium. It will extend high into the heavens and reach up among the stars and to the spheres of the circuit of the Divine Chariot, as Scripture says: "above the highest elevations of the city" (Proverbs 9:3b). The Shekinah of God and His Glory will fill its sanctuary, and there He will appoint for each angel his particular service. (Pirke MaŠia)[11]

These words do not describe the earthly city, but they do fit the New Jerusalem,

The foundations of the city walls were decorated with every kind of precious stone. The first foundation was jasper, the second sapphire, the third chalcedony, the fourth emerald, the fifth sardonyx, the sixth carnelian, the seventh chrysolite, the eighth beryl, the ninth topaz, the tenth chrysoprase, the eleventh jacinth, and the twelfth amethyst. (Revelation 21:19, 20)

They also fit the shining city of the book of Tobit 13:16b:

The gates of Jerusalem will be built with sapphire and emerald, and all your walls with precious stones. The towers of Jerusalem will be built with gold, and their battlements with pure gold. The streets of Jerusalem will be paved with ruby and with stones of Ophir.

India: The Wheel of Dharma (Righteousness)

Dharma wheel (ES)

Righteousness and justice are the foundation of your
throne; Mercy and truth go before your face.
—Psalm 89:14

The Vishnu Purana paints a striking portrait of the city of Devaraka (Sanskrit: divine, celestial) called the Matchless:

The city was square—it measured a hundred yojonas, and over all, was decked in pearls, rubies, diamonds, and other gems. The city was high … with emerald pillars, and with courtyards of rubies. It contained endless temples. It had *crossroads* decked with sapphires, and highways blazing with gems. It blazed like the meridian sun.[12]

This beautiful city departed in a great disaster when the oceans submerged it. The Mahabharata calls it Ekacakra (one wheel of dominion) and Paricakra (dominion wheel all around).[13]

The Maha Sudassana Suttana describes the heavenly city of Kusavati. Its building materials were of gold, silver, beryl, crystal, agate, coral, and other gems. Surrounded by seven ramparts, it has four gates of gold, silver, crystal, and jade. It goes on to describe the area beyond:

> Beyond the ramparts were *seven* rows of palm trees, the fourth row having trunks of silver and leaves and fruit of gold; then followed palms of beryl, with leaves and fruit of beryl; agate palms, whose fruit and leaves were of coral, and coral palms with leaves and fruit of agate; lastly the palms whose trunks were composed of all kind of gems had leaves and fruits of the same description, and when these rows of palm trees were shaken by the wind, arose a sound sweet and pleasant, charming and intoxicating. (Trans. TW Rhys Davids 1881, *Sacred Books of the East*, edited by Max Müller)

Ireland and Wales: Tir na n-Og and Avalon

Not everyone who says to me, "Lord, Lord," will enter the kingdom of heaven, but only he who does the will of my Father who is in heaven.
—Matthew 7:21

In old Irish tradition, Heaven was Roth Fáil (Wheel of Destiny),[13] and within it was Tir na n-Og (Land of Youth). It is a place of peace and harmony in the north where instruments sound without being played, from the very stones. Sweet, warm streams flow through it, the choice of mead and wine, splendid people without blemish. Brightly plumed birds sing enchanting music from the swaying branches of the Otherworld Tree.[14] Seven zones of land surround the great central hall of Tara where the High King resides.[15]

The Celtic cross resembles the Land of God (ES)

In Welsh lore, Heaven is Arionrhod (Bright Wheel), and the garden paradise within it is Avalon, a quadrangular loadstone castle in the far north. With halls lighted with self-luminous stones, every kind of jewel adorned it. In its center was a marvelous tree bearing golden apples and a sparkling fountain with four streams flowing from it.[16] A lovely description of it appears in the chronicle of *The Voyage of Bran.*

> There is a distant isle.
> Around which seahorses glisten
> A fair course on which the white wave surges.
> *Four pedestals uphold it ...*
> Pillars of white bronze beneath it
> Shining through the eons of beauty.
> Lovely land through the ages of the world
> Unknown is wailing or treachery in the happy familiar land;
> No sound there rough or harsh,
> Only sweet music striking on the ear.
> Then if one sees the silvery land
> On which dragon stones and crystals rain
> The sea breaks the wave upon the land.[17]

Greece and Rome: The Elysian Fields

Your eyes will see the King in His beauty and
view a land that stretches afar.
—Isaiah 33:17 KJV

In Greek and Roman literature, Elysium shines above the earth. Plato speaks of it as a "visible, tangible heaven."[18] In his *Vera Historia*, Greek orator Lucian speaks of emerald walls, temples of beryl, and altars of amethyst, but Elysium itself was of pure gold. Life is filled with laughter and song, and soft breezes constantly refresh it. Here, virtuous men and women rest after death. With tame animals and singing birds, it is a land of perpetual youth and continuous sunshine with all kinds of fruit always in season.[19]

China and Tibet

The kingdom of heaven is like treasure hidden in a field.
When a man found it, he hid it again, and then in his joy
went and sold all he had and bought that field.
—Matthew 13:44

The early Chinese character for Tiān (heaven) is ⊕.[20] The sages have described it as the garden of Kunlun 崑崙山 in the Land of Extreme Felicity. In this pure land, the virtuous rest on flowery carpets, listen to the melodious warbling of colorful birds, and feast on delicious fruits hanging from luxuriant groves.[21] In the center of the garden is the peach tree of immortality and a fountain with four rivers flowing to the four directions.[22]

> This happy land is surrounded with a sevenfold row of railings, a sevenfold row of silk nets, and a sevenfold row of trees. In the midst of it there are seven precious ponds, the water is still, it is pure and cold, it is sweet and agreeable, it is fresh and rich, it tranquilizes, it removes hunger and thirst, and finally it nourishes all roots. The bottom of these ponds is covered with golden sands, and round about there are pavements constructed of precious stones and metals, and many two-storied pavilions built of richly colored transparent jewels.

On the surface of the water there are beautiful lotus-flowers floating, each as large as a carriage-wheel, displaying the most dazzling colors, and dispersing the most fragrant aromas. There are also beautiful birds there, which make delicious, enchanting music, and at every breath of wind the very trees on which those birds are resting join in the chorus, shaking their leaves in trembling accords of sweetest harmony.[23]

Shambhala

Preserved in the oral traditions of Tibet is the paradise of Shambhala, a land as brilliant as the sun. It is separated from this world by a supernatural defense and inhabited by dharma kings. Shambhala's reappearance is expected after a series of global catastrophes and signs and wonders in the heavens. During this apocalypse, an evil leader will arise to conquer the earth. He will attempt to lay siege to the mystical kingdom itself. Then, the last righteous king, Rudrachakin (Powerful King Holding an Iron Wheel) will return riding a white warhorse. With a great army, he will destroy all evil forces with spiritual power and usher in an age of peace lasting a thousand years.[24]

For Tibetan Buddhists, Shambhala is an actual, though invisible, place in the north associated with the aurora borealis. Bordered by rings of glaciers, it has palaces of gem-encrusted halls and pillars of precious stones. There is no suffering, conflict, pain, or death.[25] In the tantra of the Kalachakra (Wheel of Time), the fourteenth Dalai Lama explains that if we could build spaceships capable of transporting us there, the tickets would be very expensive. They would be "meritorious actions."[26] Of course, no one can enter Heaven this way. It is a free gift. "For it is by grace you have been saved, through faith—and this is not from yourselves, it is the gift of God—not by works, so that no one can boast" (Ephesians 2:8, 9).

Japan: The Pure Land

At once I was in the spirit, and there in heaven stood
a throne, with one seated on the throne!
—Revelation 4:2

In the sacred books of Japan, the Throne of God is "the One." It is separated from this world by fire and water. In its center is an ambrosial pond with floating lotuses called the Pond of Mercy. There, beautiful flowers bloom, each one of a different hue, the colors vibrating intensely.[27] Lofty palaces and pavilions are decorated with the finest jewels. Angels play music, and colorful birds sing heavenly songs from terraces of fragrant trees. Bells hanging from the trees play sweet music in the soft, gentle breeze as hosts of angels scatter flowers. The host of heaven holds its assemblies in this pure land, Gokuraku Jōdo 極楽浄土.[28]

The Flower of God

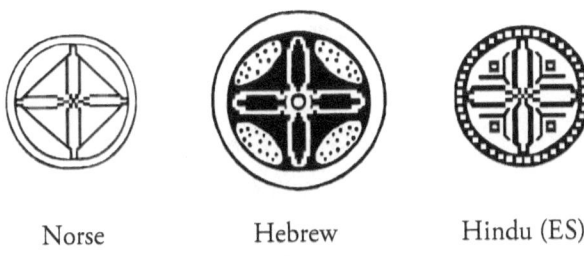

Norse   Hebrew   Hindu (ES)

I am the rose of Sharon, and the lily of the valleys.
—Song of Songs 2:1

The ancients remembered Heaven's resemblance to a beautiful flower and decorated their temple altars, tombs, and thrones with these living symbols of the dwelling place of God. Plants with four petals or leaves (dogwood, holly, four-leaf clover, evening primrose, etc.) were especially revered. The Egyptian hieroglyph Tu ⚜ (the One, the King) is called the primordial flower. It appears on the Moses chair of the ancient Synagogue of Chorazin (Matthew 23:2, 3).

The Egyptians called their beloved Asar (Osirus) "Holy Flower" and Seshen-uab (Holy Lily).[29] For four thousand years, no other god usurped his exalted position. In the early centuries, there were no images of him. "Before the XIIth dynasty figures or representations of Osiris are very rare and some doubt if any exist."[30] His glyphs are a throne and an eye 👁. He was Un Nefer (the Beautiful Being). As merciful judge of the dead, he rewarded the righteous and punished the wicked. He was "Lord of Life, Tua-neter (Morning star of God), Governor of Eternity, King of Kings, Lord of Lords."[31]

Associating rulers with flowers is traditional in Asia. The kingdom of Yú Tiān Shangdi上帝(Supreme Ruler of Imperial Heaven) is compared to a golden flower. China calls itself Zhōng Hua中花 (Central Flower).[32] The emperor's throne in Japan is still known as the Chrysanthemum Throne.

Variations of the primordial flower can be seen on temples and monuments all over Mexico, Central America and South America. "One Flower" was a popular title of God. Xochipilli is the Mayan principal flower. The Aztecs said that the flower death came down from heaven.[33] The Codex Chimalpopoca calls the God of the Flood: Nahui-xochitl (Four Flower) and Nahui-atl (Four Water).

In Israel, the rosette adorned priestly garments, temple walls, and vessels. "And you shall make sockets or rosettes of gold for settings" (Exodus 28:13 AMP). The rosette is the floral version of the Star of God. As symbols of Heaven, rosettes are seen in funerary art all over the ancient world. Here is the prototype of all the rose windows in our cathedrals and the blessing signs in our folk art.

The Star of God and Four Rivers on Mesopotamian cylinder seal.
Artwork courtesy of Ava Raha.

Throughout the world, eight is the number of eternity, infinity, and new beginnings. One and eight are the same musical tone. In Japan, the Star of God combined with the four rivers is called Ame-No-Ya-Timata (the heavenly eight crossroads).[34] Ya, one of the ancient names of God, is "eight." The Taoist Ba Gua and Buddhist eightfold path derive from this source, as does the design of the octagonal Byzantine church. Here is the origin of Odin's eight-legged horse Sleipnir. The eight theme is continued in the Greek and Hebrew, languages in which letters also stand for numbers. The number of Jesus is 888, and the Father is 777.

As symbols of divine purity, lilies adorned the pillars of ancient temples. "On the tops of the pillars was lily-work" (1 Kings. 7:22). In India, the water lily (golden lotus) symbolizes Brahman's heaven resting on the primeval waters. The white lotus and blue lotus are sacred throughout Asia. *The Lotus Sutra* calls Heaven Meghapushpa (cloud blossom). Shambhala resembles an eight-petaled lotus.[35] Millions daily pray, "Om Mani Padme Hum" (Hail to the Jewel in the Lotus) while turning their prayer wheels. The lotus position, sitting immobile with legs crossed, imitates the sign of God in the flesh.

The Great Weaver

And he made for the altar a brazen grate of network under the compass.
—Exodus 38:4 (KJV)

Recently, certain astrophysicists suggested that the fabric of the universe is woven together with myriads of vibrating strings. The ancients were familiar with this string theory. They saw their god as the spinner and weaver of a vast cosmic fabric. The rabbis describe the branching streams of the heavenly river as the thread of the warp ‖ and the woof =. The Even ha-Shetiyah is the Foundation Stone from which the world was created (Isaiah 28:16). Shetiyah in Aramaic means "weaving" or "to fix the warp."[36]

The Rg Veda of India also speaks of this: "For both the warp and the woof he understandeth, and in due time shall speak what should be spoken."[37] Brahman is "He upon whom the worlds are woven as warp and woof."[38] The *Nihongi* of Japan calls heaven "a weaving hall."[39] Blankets, carpets, and garments are regaled with the sign of God. The Scots still proudly wear His tartan, and His sign can be seen on funerary art worldwide.

The Kingdom of Heaven Is Like a Net

Moreover, no one knows when their hour will come: As fish are caught in a cruel net, or birds are taken in a snare, so people are trapped by evil times that fall unexpectedly upon them.
—Ecclesiastes 9:12

Closely related to the web is the net. The Hebrew first word of Genesis, BíReshit ("In the beginning"), derives from the word ReShet (a woven net). According to Exodus 27:4, 5, the temple altar had a bronze net. Many cultures see the Creator as a great "Fisher-King" catching souls in His net. The four winds are associated with a net in the Babylonian Second Tablet of creation:

> The south wind, the north wind, the east wind, and the west wind he held the net close to his side, the gift of his father Anu.[40]

In Micronesian tradition, a divine fish trap was lost through disobedience. Men and women were innocent in the beginning. There was a tree that Nakaa, the guardian, forbade them to touch, but they disobeyed him. Noticing the men's gray hairs, he said, "Fools, death has come to you." He took away the fish trap and tree of life and left them. Then he began weaving a net for catching souls. Wrongdoers were eternally entangled, but good people joined their ancestors.[41]

There are many biblical allusions to a net. Jesus said:

> Once again, the kingdom of heaven is like a net that was let down into the lake and caught all kinds of fish. When it was full, the fishermen pulled it up on the shore. Then they sat down and collected the good fish in baskets, but threw the bad away. This is how it will be at the end of the age. The angels will come and separate the wicked from the righteous and throw them into the fiery furnace, where there will be weeping and gnashing of teeth. (Matthew 13:47–50)

> Be careful, or your hearts will be weighed down with dissipation, drunkenness, and the anxieties of life, and that day will close on you unexpectedly like a trap. For it will come upon all those who live on the face of the whole earth.
> —Luke 21:34, 35

Celestial Idolatry

For all the gods of the peoples are idols: but the Lord made the heavens.
—Psalm 96:5 (NKJV)

Because the kingdom of God was once visible, there was constant temptation to portray it. That is why graven images of celestial things are the first thing forbidden in the second commandment (Exodus 20:4). At first, the people differentiated between God and His dwelling place by drawing stars, crosses, wheels, flowers, and so on. But as memories of His kingdom faded, there arose confusion between the King and His throne:

> Professing to be wise, they became fools, and changed the glory of the incorruptible God into an image made like corruptible man—and birds and four-footed animals and creeping things. (Romans 1:22, 23 NKJV)

The Bible always differentiates between the Most High God and His dwelling place. When idols are constructed, the connection with the living God ceases. Demons, who desire worship and a house, occupy idols and force humans to do their bidding. "The idols speak deceitfully, diviners see visions that lie; they tell dreams that are false, they give comfort in vain. Therefore the people wander like sheep oppressed for lack of a shepherd" (Zechariah 10:2). Our ancestors assert that these are spirits of an antediluvian human-hating race.[42]

Ganesh, Hindu God of beginnings, wisdom, and crossroads.
Artwork courtesy of Ava Raha.

The sign of God is seen on many tribal idols sitting immobile with legs crossed. The Chinese word Fu 父 (father, to sit cross-legged) also means "to be subject, to fall prostrate, to remove evil."[43] Primitive imagery accounts for the grotesque appearance of many idols with four faces, four limbs, four teeth, and so on. Over the centuries, some of the more vulgar elements were replaced as they became increasingly anthropomorphic.

The Aztecs were widely known for their idolatry. By the time the Spanish arrived, their religion was in deep decay through centuries of obedience to idols. A powerful priesthood, who used hallucinogenic drugs, subjugated the people through fear.[44] They learned human sacrifice from visions shown to them by their idol, Huitzilopochtli.[45] He demanded human blood and hearts, the only worthy offering to become nourishment for the great star that maintains life on earth.[46] Since hearts were obtained from war prisoners, peace was dangerous, and "flower wars" were fought to provide sacrificial victims.[47] The victim was spread-eagled like a chi cross on a swaddled stone altar. The victim actually became the god.

The Quiché Mayans were enslaved by an image called Tohil, who demanded human sacrifice.[48] At his command, they packed their idols on their backs and began a long migration. Constantly watching for the appearance of the morning star, they sang a song called "The Blame Is Ours." They sacrificed their blood to Tohil by passing thorn-embedded cords through their ears, elbows, and tongues.[49]

The Weeping Viracocha
—HB Alexander, "Latin American Mythology" Vol. 11
Mythology of All Races (1916)

The Hopi of the American Southwest belong to the Shoshonean branch of the Uto-Aztecan language family. They say that the Mayans and Aztecs, with whom they shared many common traditions, are "aberrant Hopi clans" who perished because they failed to follow the Creator's ordained plan.[50] The people of the Andes once said of their Creator, Viracocha, "It was forbidden to represent him in any form."[51] But later, he was sculpted with crosses, a tiara, and a long, flowing garment. Idolatry led to their enslavement. Before their banishment, their idols had greatly harmed the people, forcing them to make human sacrifices to them.[52]

Some nations did not represent their gods with idols. Traditional Japanese worship, Shinto Kami no Michi (the Way of the Gods), does not permit the use of graven images:

> Never make an image in order to represent the Deity …
> If we try to establish a relation between the Deity and
> man indirectly by means of an image, the image will itself
> stand in the way and prevent us from realizing our religious
> purpose to accomplish direct communion with the Deity,
> so an image made by mortal hands is of no use in Shinto
> worship.
>
> —Ise-Teijo (1715–1784)[53]

Although the prophets constantly warned against it, Israel practiced idolatry through apostasy and intermarriage with neighboring tribes (Deuteronomy 29:24–28). The people worshipped images of a star-God: "[No] but [instead of bringing Me the appointed sacrifices] you carried about the tent of your king Sakkuth and Kaiwan [names for the gods of the planet Saturn], your images of your star-God which you made for yourselves [and you will do so again]" (Amos 5:26 AMP); see also Acts 7:41–43). "At every crossway you built your high place [for idol worship] and have made your beauty an abomination …" (Ezekiel 16:25 AMP). There was only one way to deliver Israel and all humanity. The Lord of the universe, through a righteous remnant of Abraham's seed (Ezekiel 14:22), clothed himself in flesh and blood and came down Himself!

Thy Kingdom Come!

God reigns over the nations; God sits on his holy throne.
—Psalm 47:8 (NKJV)

Plato (428–348 BC) said that the disorder inherent in nature did not exist when the true shepherd reigned and that the best ordered of existing states is only an inferior copy of His kingdom. When men rejected His righteous rule, "the Pilot let the helm go and retired," and men were then left to their own devices.[54]

The natives of the Ivory Coast, Ghana, Togo, and Nigeria speak of a past age when God lived near humans directly overhead. But He became angry about their behavior and retired from the earth, and none have seen Him since.[55] Indigenous Australians look back to the dreamtime when the visible supernatural beings disappeared from the sky.[56] Native Hawaiians say that during the Era-of-over-turning (Po-au-hulihia) in the time of Nu'u, came the flood known as Kahinali'i (the Sea that made the chiefs *(ali'i)* fall down).[57] They call Paradise Kahiki-honua-kele (the land that moved off).[58]

The Chinese word tiān 天 means "high God, heaven,"[59] but it also means, "exterminate." The ancient sages describe a cosmic dissolution that separated the higher from the lower ages of humanity,

> The pillars of heaven (天柱 tiānzhù) were broken; the earth shook to its very foundations; the heavens sunk lower toward the north; the sun, the moon, and the stars changed their motions; the earth fell to pieces, and the waters enclosed within its bosom, burst forth with violence and overflowed it. Man having rebelled against heaven, the system of the universe was totally disordered. The sun was eclipsed, the planets altered their course, and the grand harmony of nature was disturbed. (Sir William Jones, *Asiatic Researches*, volume 2)

In other parts of Southeast Asia, it is said, "The sky was inverted and moved away from earth."[60]

Gotland Rune stone, Sweden (ES)

In the Old Norse accounts, human beings were abandoned when the world of the gods, ruled by the great All-father Odin, vanished during Ragnarök, the Old Norse name for the Flood, from ragna (gods) and rök (judgment). The sun was dark at noon, the celestial lights disappeared, the earth shook, and the "terrible winter" came.[61]

According to many African tribes, it was God dwelling among humans that made the world before the Flood so wonderful. Others blame human depravity for the disappearance of God into the sky.[62] The Yao, one of the Bantu tribes of Mozambique, ascribe His disappearance to humans' destruction of the harmony of the natural world.[63] The Bushmen of the Kalahari Desert say the Creator Mulungu once lived with human beings, but their foolishness drove him far away, and now many of us are hungry.[64]

Our ancient forebears expected their ancestral Father's return to restore all things to their original perfect state,

> And I heard a loud voice from the throne saying, 'Now the dwelling of God is with men, and he will live with them. They will be his people, and God himself will be with them and be their God. He will wipe every tear from their eyes. There will be no more death or mourning or crying or pain, for the old order of things has passed away. (Revelation 21:3–4)

Chapter 3

The Chosen Place of God

The Unmoved Mover at the Center

Yet God is my King of old, working salvation in the midst of the earth.
—Psalm 74:12 (KJV)

The Almighty is not limited to any location, but it is evident from His Word that He has chosen four places—the First Place, the Central Place, the Highest Place, and the Last Place. Each of these has a rich mythology, but all four are the very same place. During the age between the fall and the Flood, our ancient ancestors located the Throne of God at the heart of the sky, the position of the present sun at noon (Psalm 37:6). Although everything moved about Him, His kingdom remained steadfast and immovable directly overhead. "Every good gift and every perfect gift is from above, and comes down from the Father of lights, with whom is no variation, nor shadow of turning" (James 1:17 NKJV).

We see the importance of the middle or central place in the Bible. The tree of life "is in the *midst* of the Garden" (Genesis 2:9; see Revelation 22:2). "It shall be the holy district, and the sanctuary of the temple shall be in the *center*" (Ezekiel 48:21b). Solomon consecrated the *middle* part of the courtyard in front of the temple of the Lord and there he offered burnt offerings" (2 Chronicles 7:7a). "The sun stopped in the *middle* of the sky and delayed going down a full day" (Joshua 10:13b).

The feasts of Passover and Unleavened Bread begin in the *middle* of the *first* month Nissan (Leviticus 23:5), the time of the first redemption of Israel from bondage in Egypt. It is believed that the second redemption of Israel will occur at the same time. The Feast of Tabernacles begins on the fifteenth day of the seventh month (Leviticus 23:39), the *middle* of the year. The number fifteen is the numerical value of Yah, one of the names of God. Purim is celebrated in the middle of the month of Adar.

The classical writers knew the Creator as the Unmoved Mover in the center of the sky:

> In every case where objects are moved, there is an original Unmoved Mover.
> —Aristotle, *Natural Science*

The Greek philosopher and poet *Xenophanes* proclaimed:

> There is one God, greatest among gods and men, neither in shape nor in thought like unto mortals. He is all sight, all mind, all ear ... *He abides ever in the same place motionless, and it befits him not to wander hither and thither.* (Fragment, Xenophanes of Kolophon 536 BC)

Dante, well acquainted with the Bible and the ancient classics, used the same language in his *Divine Comedy*:

> I believe in one God, sole and eternal, who, unmoved, moves all the heavens with love and with desire; and for this belief I have not only proofs physical and metaphysical, but it is given to me also in the truth that rains down hence through Moses and the Prophets and the Psalms, through the Gospel. (*Paradiso* Canto 24)

Thomas Campbell was familiar with this truth as seen in this poem:

> The shrine where motion first began,
> And light and life in mingling torrent ran,
> From whence each bright rotundity was hurled,
> The Throne of God, the Center of the World.
> (Thomas Campbell, *Pleasures of Hope* 1799)

The Egyptians spoke of the One God (Ua Neter) who rests in the middle of the sky. In hymns, he is addressed as "Thou shining one in the kingdom, thou stable one in the place of stability."[1] The Supreme God of ancient Japan is at rest at the center of the cosmos yet moved all things. Among his titles are Ame-no-mi-naka-nusi-no-Kami (God-at-the-Very-Center-of-Heaven), Ame-no-toko-tachi-Kami (God-Eternal-Stand-of-the-Heavens), and Kuni-no-toko-tachi (The-One-Who-Stands-Perpetually-Over-the-World).[2] His eternal Rock-seat, Ame-no-Iha Ya, is near the heavens-river (Kojiki 1. 16, 32).

The identical concepts are found in the sacred texts of India.

> Who but the purest of the pure can realize this effulgent being
> who is joy and beyond joy. Formless is he, though inhabiting form.
> In the midst of the fleeting he abides forever.
> —Katha Upanishad (trans. Sir Monier-Williams)

> There is one only being who exists
> Unmoved, yet moving swifter than the mind;
> Who far outstrips the senses, though as gods
> They strive to reach him, who, himself at rest,
> Transcends the fleetest flight of other beings;
> Who, like the air, supports all vital action.
> He moves, yet moves not.
> —Isha Upanishad (trans. Sir Monier-Williams)

> He standeth in the dwelling of the Highest, a Pillar
> on sure ground *where the paths are parted.*
> —*Rg Veda,* book 10, Hymn 6
> Ralph TH Griffith, trans. *Hymns of the Rg Veda* (1889)

In the Mahabharata, he is "the one in the sky who suffuses the vault of heaven with benevolent splendor."[3]

The hours of noon and midnight were sacred to all ancient peoples. It was at noon when the Lord was crucified (Luke 23:44–46). At noon, St. Paul saw the great light on the road to Damascus (Acts 26:13). At noon, St. Peter was shown that the gospel was also for the Gentiles (Acts 10:9). The middle of the month was auspicious. The Flood began on the fifteenth day of the month Daisios.[4] The ides were sacred to the Romans as they were the middle of the month (the fifteenth of March, May, July, October; the thirteenth in the others). The word derives from the Etruscan *iduare* (division). The Norse All-father Odin ruled from his high throne in Midgard. For him (Anglo-Saxon Woden*)*, Wednesday is named—the middle day of the week.[5]

The great Egyptian festivals were celebrated on the first and the fifteenth days of the month. The word for the middle of anything is *met-t* or *meti*. *Meter-t* means "noon, midday." *Met er* is "between" (Greek *meta;* Sanskrit *madhya)*. *Meten* is "the way or path of heaven." *Metit* or *metrit* is "righteousness, integrity." *Metu* is "right order." Metut Neter are hieroglyphs, the "words of the god." *Met-tua* means "fifteen," and the festival of the fifteenth day of the month.[6]

The Navel of Heaven

For the Lamb at the center of the throne will be their
shepherd and he will lead them to springs of living water.
And God will wipe away every tear from their eyes.
—Revelation 7:17

The ancestral God of all ancient tribes was called "The Navel of the Sky," a position represented by the universal sign ⊙, the central dot depicting the Creator's holy seat at the pole of the universe.[7] The Hindu Brahman is Nabhi-ja (navel-born). The Vedic Agni of the fire altar is called Nabhas (navel). The Assyrian Creator God Asshur is the nave of the universal wheel.[8] The supreme God of the ancient Finns, Ukko, whose title, Taivahan Napanen (Navel of Heaven), is recalled in song:

Where has fire been cradled, where rocked the flame? Over there
on the navel of the sky, on the peak of the famous mountain.
—Uno Holmberg, "Finno-Ugric Mythology,"
Vol. 4 *Mythology of All Races* (1916)

Since their god lived at the navel of the sky, all tribes believed their cities were directly under heaven; Babylon, Athens, Delphi, Paphos, Samarkand, and so on. Qebhu Nippur (navel), the oldest city of Sumer, was said to be the first of five cities founded before the Flood.[9] The Collao people of the central Andes say their village is the middle of the world from which the men who repopulated the earth after the Flood set forth.[10] The natives of Easter Island called their homeland Te-Pito-o-te-Henua (Navel of the World).[11] The Chickasaw believed Mississippi to be the center of the earth, and the mounds of their country were called "navels."[12]

As the navel is the connection to the womb, our forefathers saw that body part as symbolic of their connection to God. Archaic customs including *omphaloskepsis* (the mystical contemplation of the navel) survive to this day. Words associated with *middle*—meditation, model, medicine (from the Latin *medere* to heal)—have celestial origins, as do the names Medes, Medius, Medea, Medina, Medbh, Midian, Dromedes, Andromeda, and so forth.[13] The name Mediterranean means "middle earth."

The word *navel* itself has an interesting ancestry. The English *nave* (Old English *nafela;* Old Teutonic *nabalon*) means "the hub or central part of a wheel from which spokes radiate" and "the central part of a church."[14] The similar sounding Hebrew word *nâvâh* is "to be at rest in a beautiful dwelling place." A prophet, who stood on the navel-stone of a consecrated circle to proclaim the Word of God, was a *navi* (from *nava* "to prophecy"). Plato wrote:

> These are matters of which we are ignorant ourselves, and as founders of a city we should be unwise in trusting them to any interpreter but our ancestral deity. *He is the god who sits in the center, on the navel of the earth,* and he is the interpreter of religion to all mankind—Plato, *The Republic* Bk. 4 (trans. Benjamin Jowett)

The Middle Kingdom

Have we not all one Father? Did not one God create us?
—Malachi 2:10a

Yú Huang Tiān Shangdi 天玉上帝 (Jade-Sovereign-Heaven-Rule)[2] was the first ancestor and founder of the Chinese Empire.[15] His "entire countenance was super-eminently beautiful, so that none became weary in beholding him."[16] He exercised "universal benevolence," saving the people.[17] Devotion to Shangdi 上帝(or Shang Ti) was the most sacred form of worship. In the

---

[2] His name is often shortened to Shang-Ti, Shangdi 上帝, Huang-Ti, Huangdi 皇帝, Tiāndì 天帝(heaven-god), or Yùti (jade, glorious god).

archaic Shu King, compiled by Confucius, the *Ming* 命運 (celestial destiny) of princes and individual men is attributed to the action of Shangdi. Confucius declared, "The ceremonies of the sacrifices to heaven and earth are those by which we serve Shangdi."[18] He taught the people "how to control the forces of nature and their own hearts."[19]

The earliest character for Shangdi 奭 incorporates the land of God.[20] He was the embodiment of the Tao (the Way of Heaven) and "represented the Great Law to which everything in the universe is subject."[21] The sages called the Tao "the Originator of All Things, the Root and Ground of Existence, Maker and Transformer, the Center of Attraction for the Whole World."[22] The Chinese, like other ancient peoples, saw their land as the only true reflection of Heaven. China calls itself Zhōng guó (the Middle Kingdom).[23] Zhōng 中 is "middle, center." The sage Lao Tzu called the present sky "the Empty Heart," since "Huang-Ti withdrew and gave up his government."[24]

Face-to-Face

In the light of the king's face is life, and his
favor is like a cloud of the latter rain.
—Proverbs 16:15

In rabbinic tradition, the lower world and the upper world were once face-to-face. Heaven and earth were created as a unit, although made of different elements.[25] Biblical phrases, such as "the Face of Heaven" and "the face of the deep," survived the Flood. The earth also had a "face" (Exodus 32:12; 33:16). One of the oldest Egyptian names for God was Heru, "the Face of Heaven."[26] In the Pyramid Texts, Heru (Horus) is called the morning star. Four scepters or pillars supported his dwelling at the four cardinal points.[27] Ptah, supreme god of Memphis, "who stretched out the heavens," was known as "the god of the beautiful face."[28]

The Japanese chronicle, The Kojiki, speaks of the deity Omo-daru no-kami (perfect-face or perfectly beautiful).[29] The Mayan god worshipped in the Yucatan was known as Kinich-ahau (Lord of the Face of the Sun). He presided over the north.[30] All the earliest origin accounts describe a heaven-centric cosmos. During this unique era, from Adam to Noah, all human beings faced their God at the zenith of the sky. The earth was a satellite of the star of God and was astroasynchronous. In other words, it always turned the same face to its primary, like the relationship between the earth and the moon today.[31]

Recalling that sacred time, the Egyptians prayed:

> Turn thou to us thy face, O Sovereign our Lord! Life is [on] our fac[es] in the seeing of thy face, turn not thou away thy face from us. The joy of our heart is in the sight of thee, O beautiful Sovereign, our hearts would see thee. (*The Lamentations of Isis and Nepthys*)

This prayer is very similar to, "Turn us again, O God, and cause thy face to shine; and we shall be saved" (Psalm 80:3 KJV) and "Turn us back to you, O Lord, and we will be restored; renew our days as of old" (Lamentations 5:21a NKJV). Why did God hide His face? "For you have hidden your face from us, and have given us over to our sins" (Isaiah 64:7b). The Bible describes the cosmos in its divinely ordered state, rather than its present temporary situation.

The holy place of the tabernacle of the Most High. God is in the midst of her; she shall not be moved … (Psalm 46:4b, 5a NKJV).

The world also is established that it cannot be moved. Thy throne is established of old (Psalm 93:1, 2 KJV).

Say among the nations, 'The Lord reigns.' The world is *firmly established*, it *cannot be moved* (Psalm 96:10a).

Wherefore we receiving a kingdom which *cannot be moved*, let us have grace, whereby we may serve God acceptably with reverence and godly fear: for our God is a consuming fire (Hebrews 12:28, 29 KJV).

For now we see only a reflection as in a mirror; then we shall see face-to-face. Now I know in part; then I shall know fully, even as I am fully known (1 Corinthians 13:12).

When the heavens change again, we will see Him face-to-face. "He holds back the face of his throne, and spreads his cloud upon it. He has compassed the waters with bounds, until the day and night come to an end" (Job 26:10). This is the prayer of the righteous of all ages, "Be exalted, O God, above the heavens. Let your glory be over all the earth" (Psalm 57:5). "You have set your glory above the heavens" (Psalm 8:1b). "'All mankind will come and bow down before me,' says the Lord" (Isaiah 66:23b). "All the families of the nations will bow down before him, for dominion belongs to the Lord and he rules over the nations" (Psalm 22:27b). "The glory of the Lord will be revealed, and all mankind together will see it" (Isaiah 40:4, 5).

Norse runic shaman drum
Uno Holmberg, Vol. 4 "Finno-Ugric Mythology" *Mythology of All Races* (1916)

The Upright Axis

The Lord is upright; he is my rock, and there is no
unrighteousness in him. —Psalm 92:15

Unlike the animals whose backbone is parallel to the earth, God created the human axis to be upright. He also designed the earth's axis to be upright, so that it always pointed to Him. This made the climate uniform throughout the year. The present awkward 23.45-degree inclination of the axis is a consequence of the Flood. The Greek philosopher Anaxagoras (c. 500–428 BC) described this basic tenet of our early ancestors,

In the beginning the stars moved in the sky as in a revolving dome, so that the celestial pole, which is always visible was vertically overhead; but subsequently the pole took its inclined position.[32]

The book of Enoch, quoted in Jude 1:14, comments on this inclination:

In those days Noah saw that the earth had become inclined, and that destruction approached. Then he lifted up his feet and went to the ends of the earth to the dwelling of his great grandfather, Enoch. And Noah cried with a bitter voice: Hear me; Hear me; Hear me; three times. And he said, Tell me what is transacting upon earth; for the earth labors, and is violently shaken. Surely I shall perish with it. After this there was a great perturbation on earth, and a voice was heard from heaven. I fell down on my face when my great grandfather, Enoch came and stood by me.[33]

Thomas Burnet wrote in *Sacred Theory of the Earth* (1681):

The Poles of the World did once change their situation, and were at first in another posture from what they are in now, till that inclination happen'd [...] the earth chang'd its posture at the deluge, and thereby made these seeming changes in the heavens; its Poles before pointed to the Poles of the ecliptic which now point to the Poles of the equator, and its Axis is become parallel with that Axis [...] at first they say, there was no variety of seasons in the year, as in their Golden Age.[34]

John Milton wrote of this ancient displacement in *Paradise Lost*:

Some say he bid his angels turne ascance
The poles of earth twice ten degrees and more
From the sun's axle;' they with labor push'd
Oblique the centric globe; some say, the sun
Was bid turn reins from the equinoctial road ...
To bring in change of seasons to each clime;

Else had the spring
Perpetual smil'd on earth with verdant flowers,
Equal in days and nights except to those
Beyond the polar circles; to them day
Had unbenighted shone; while the low sun
To recompense his distance, in their sight
Had rounded still the horizon, and not known
Of east or west. —*Paradise Lost* (1665)

The ancient Persian Zoroastrian texts (Zoroaster means "seed-star") describe the cosmos in its perfect state when the celestial lights stood still over the earth and "the Great One" was "in the middle of the sky."[35] This changed when the raging Angra Mainyu (the devil as a serpent) sprang up through the sky at noon, plunged down into the waters, and burst through the center of the earth, causing the mountains to grow. He assaulted the water and desiccated the plants. The heavenly bodies were shaken from their places and began to move. The fixed order of the sky was destroyed. There was smoke and fire, and noon became as dark as night. A future restoration and final defeat of evil is predicted. The resurrection will occur at noon called "the time of Rapithwin."[36]

The Egyptian hieroglyph for *neter* (god) is a flagstaff. This glyph conveys His uprightness and the original position of the earth's axis. The related word Soter ⌐ signifies "Savior-god."[37] Another glyph for Neter, the seated god, is an illustration of Psalm 29:10 (NKJV), which states, "The Lord *sits enthroned at the flood*, and the Lord sits as King forever." Like the Psalmist, the One God (Ua Neter) in every *nome* (province) of Egypt, is represented in a seated upright position.

The ancient texts of India speak of the One who sat above. In the Hymn to Na'ra'yena (moving on the waters), He is love, goodness, light.

Thou sat'st alone; till, through thy mystic Love
Things unexisting to existence sprung,
And graceful descant sung.
What first impelled thee to exert thy might?
Goodness unlimited. What glorious light.
(Trans. by Sir William Jones)

The Perfect Year

The Lord looked down from his sanctuary on
high, from heaven he viewed the earth.
—Psalm 102:19

The upright axis produced a perfect year of 360 days now called the prophetic year. The flood narrative is based on it. The Inca calendar followed this pattern. Like the Babylonians, they had the 360-degree circle, sixty-second minute, and sixty-minute hour.[38] The original year of India was 360 days.[39] March began the old Roman 360-day year of ten months (thirty-six days each).[40] The original Chinese calendar was 360 days with twelve months of thirty days. The 360-day calendar existed among the Egyptians, Hindus, Persians, Assyrians, and Greeks. Herodotus, Plutarch, Strabo, Plato, Berossus, Diodorus, and Homer all assert this. Five "unlucky," unnamed days had to be added in the seventh century BC, resulting in a 365-day year as a result of other disasters. It was later adjusted to 365 and a quarter days. The Chinese added five and a quarter days called the Khe-ying.[41]

To record linear time, the classic Mayans used a calendar called the long count with 360 days, the perfect year of Genesis. This year, called a *tun,* was divided into twenty months of eighteen days. The word tun means "stone" and "birth" in the Nahuatl and Lakota languages. Stones were inscribed and erected at the completion of each 360–day period. The base or zero date of "the long count" was August 13, 3114 BC. From this cryptic date roughly five thousand years ago, astronomer-priests began all Mayan chronology. For the highest number in the long count to change, 144,000 days were required.[42]

Mayan scholars Linda Schele and Mary Ellen Miller explain that the zero date was a "membrane" that separated two very different kinds of time and space like matter and antimatter. Mayan rituals, like those of other early tribes, attempted to bring the sacred symmetry of the former cosmos into historical time and space. Their zero date pierced that membrane. Like the Hebrews, Hindus, and Persians, the Mayans did not discard their old calendars but used them to keep track of anticipated catastrophes.[43] The Mayans call the present sun Cabrakan (Earthquake Sun) because they believe this age will end in a great earthquake.[44]

The Aztec "four-movement" glyph (ES)

Naui olin (four movement) is the Aztec day sign. Olin means, "It moves."[45] It was the birthday of the present sun and the beginning of earth's rotation. This day was celebrated with great solemnity. It was on this day that the Sun-God-of-the-Number-Four reset the order of the universe.[46] This act involved the earth's displacement and the tilting of its axis.[47] The Mayans concurred. In their creation text, Popol Vuh, the final event in the lives of the gods was the rising of the sun.

To restore the earth's axis to its original position will require the ultimate earthquake. According to Revelation, this will be "a violent earthquake, such as had not occurred since people were upon the earth, so violent was that earthquake" (Revelation 16:18). This will culminate in the arrival of the desired of all nations: "This is what the Lord Almighty says: 'In a little while I will once more shake the heavens and the earth, the sea and the dry land. I will shake all nations, and the desired of all nations will come'" (Haggai 2:7a).

Gloria in Excelsis Deo!

For this is what the high and exalted One says-he who lives
forever, whose name is holy: "I live in a high and holy place,
but also with the one who is contrite and lowly in spirit."
—Isaiah 57:15a

Since the Flood, the east, where the present sun rises, has been the direction of light and life and focus of ritual. The word *orientation* means "east," but then why did *all* ancient cosmologies locate the gate of Heaven at the North Pole? The kingdom of El Elyon, "the Most High," was physically the highest point, the apex of the sky. Because of this, His kingdom was once our polestar! The north celestial pole is the first place, the highest place in the middle of the sky. Egyptologist Heinrich Brugsch (1827–1894) described the ancient Egyptian belief that the earth joined the sky at its northernmost point. The rabbis believe the Throne of Glory is at the highest point of the universe.[48] One of the great Sephardic sages, Rabbi Yaakov Culi, proclaimed:

> If any person says that he is a god because of his wealth and power, and there were actually people who made such claims as we find in the prophets, we say of him, "If your claim is true, then complete the North. (*The Torah Anthology*, Vol. 1, 96)

Scripture confirms the sanctity of the north. "Out of the north he comes in golden splendor; God comes in awesome majesty" (Job 37:22). "When he is at work in the north, I do not see him; when he turns to the south, I catch no glimpse of him. But he knows the way that I take; when he has tested me, I will come forth as gold" (Job 23:9, 10). "Beautiful in elevation, the joy of the whole earth, Is Mount Zion on the sides of the north, the city of the great King." (Psalm 48:2 NKJV). The temple was on the north side of Jerusalem, and priests presiding at the altar faced north. Unblemished animals were offered at the north side "before the Lord" (Leviticus 1:11, 2 Kings 16:14).

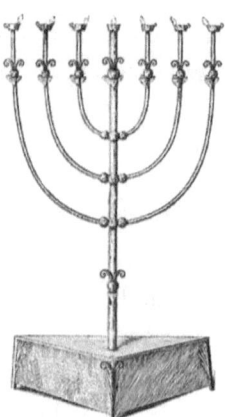

"Menorah" Jewishencyclopedia.com (1906).

The central lamp of the menorah is called the Ner Elohim (Lamp of God) (1 Samuel 3:3). The Lamp of God faced north, while all the other lamps turned toward it. This lamp burned all day, was refilled in the evening, and served to light all the other lamps. It contained no more oil than the others, yet by a miracle, the lamp regularly burned till the following evening (Jerusalem Talmud 86b). This miracle ceased after the death of Simeon the Righteous, who was high priest forty years before the destruction of the temple.[49]

The central Servant Light represents Jesus Christ. "Rather, he made himself nothing by taking the very nature of a servant, being made in human likeness" (Philippians 2:7). It is said in the Jerusalem Talmud, "Our Rabbis taught: During the last forty years before the destruction of the Temple the lot ['For the Lord'] did not come up in the right hand; nor did the crimson-colored strap become white; nor did the western light[3] shine" (Yoma 39:b).

The Etruscans recognized the north as the place from which their celestial orientation began.[50] The door of the temple faced south as the deity's place was at the north end. The Lord sat at the north end of the table facing south in the medieval halls of northern Europe, and the Master's Table at Trinity College, Cambridge, is at the north end of a north-south oriented hall to this day.[51] The heads of those buried around Stonehenge face north. Many ancient deities and their priests (e.g., the Roman Flamen Dialis, priest of Jupiter) wore the apex cap.

Flamen Dialis
Dictionary of Greek and Roman Antiquities, William Smith, LLD.

---

[3] The Lamp of God is also called the Western lamp because the Star of God came out in full glory and splendor after sunset. The present sun was not seen before the Flood: "… I will not give my glory unto another" (Isaiah 48:11b).

To the Egyptians, the north was the direction of ritual. Heads of deceased pharaohs faced north, the direction of the Neter Ta. Asar (Osiris) is addressed as Neb-er-tcher (the Lord above All) and Neb-ua (the Only).[52] Egyptologist RT Rundle Clark writes that the Egyptians, more than any other people, held the celestial pole in great sanctity. They revered a circle of stars around a central point in the north. Here was the center of order and regulation. The celestial pole is "that place, that great city."[53] A Nile poet wrote:

> There is no building that can contain him!
> There is no counselor in thy heart!
> Thy youth delight in thee, thy children;
> Thou directest them as King.
> Thy law is established in the whole land,
> In the presence of thy servants in the North ...
> He createth all works therein,
> All writings, all sacred words,
> All his implements in the North.
> (AH Sayce, *Records of the Past*, volume 4)

The Japanese deity called Ama-no-minaka-nushi-no-kami (Deity Master-of-the-August-Center-of-Heaven) and Ame-no-toko-tachi-no-kami (Deity-Standing-Eternally-in Heaven) was identified with the polestar. Vestiges of this belief remain at Ikegami (Living god), where the polestar is honored under the title Miyau Ken in the form of a wheel resting on folded hands. The Japanese, like so many others, called their land "the Middle Kingdom."[54]

The early Chinese texts located the throne of the Celestial Emperor at 北极星 Běijíxīng (the Polestar), which was Tiān-Zhōng-Gong (Heaven-Center-Palace). North was the direction of the ancestors. The emperor and his assistants, when officiating before the altar of heaven, faced north.[55] Confucius wrote:

> He who exercises government by means of his virtue may be
> compared to the North polar star, which keeps its place and
> all the stars turn toward it. (*Analects,* book 2:1)

Star in Chinese is *xīng* 星. Two names for the star of God are *kuíxīng* 魁星 (chief, best star) and *lùxīng* 禄星 (Blessing star).

The north is the most revered direction in America. The ancient Pawnee tribes linked their benevolent Creator to the North Star. Their astronomers meticulously watched the planets and their movements to ascertain the best time to make their offerings:[56]

> The Creator told them He was about to create people "like myself" … To the Star-That-Does-Not-Move he appointed the north as station, and he made him the Star-chief of the skies.
> —HB Alexander, "North American Mythology,"
> Vol. 10 *Mythology of All Races*

The Mayans regarded the north as the place housing their ancestors. They believed the dead go on to defeat death and rise from the underworld. Their spirits dwell around the North Star.[57]

The early inhabitants of India venerated the north as the direction of Brahman's heavenly city, and its sanctity is evident in texts and prayers.

> He who knows thee (the pole-star) as the firm, immovable Brahman with its children and with its grand-children, with such man children and grandchildren will firmly dwell, servants and pupils, garments and woolen blankets, bronze and gold, wives and kings, food, safety, long life, glory, renown, splendor, strength, holy lustre, and the enjoyment of food. May all these things firmly and immovable swell with me! (Grihya-Sûtras trans. by H Oldenberg)

Worshippers of all the old tribal sky gods all turned to the north in prayer.[58] In the Satapatha Brahmana II, Dhruva is Grahadhara (polestar, support), the supreme impeller in the north from which the earth sprang. On Dhruva (polestar) rests the sun. In the *Monier-Williams Sanskrit-English Dictionary*, Dhruva has thirty-seven definitions, among them "the fixed point, the enduring sound, firm, immovable, un-changeable, constant, lasting, permanent, eternal, house, pillar, mountain." In the Persian Zend Avesta, the heaven of the Creator Ahura Mazda is also in the north. The Manicheans declared, "The world (was) left to itself by the Ornament of Splendor that sustained it in the North."[59]

The German and Scandinavian temples were oriented north-south.[60] The people turned toward Asgard in the north in prayer and sacrifice:

Then the sons of Bor built in the middle of the universe the city called Asgard, where dwell the gods and their kindred, and from that abode work out so many wondrous things both on the earth and in the heavens above it. There is in that city a place called Hlidskjálf, and when Odin is seated there on his lofty throne he sees over the whole world, discerns all the actions of men, and comprehends whatever he contemplates. (*The Prose Edda* trans. IA Blackwell)

The ancient Greeks faced north to pray, because "the Father of all gods and men" was enthroned at the North Pole.[61] His palace was the Acropolis, the apex-city; his cosmos was the *polis*, the Greek city-state. From *polis*, we derive the words *politics*, *police*, and going to the *polls*. Titles, such as Zeus Hypatos (the Most High), Zeus Epakrios (the Extreme), and Zeus Polios, all refer to his heavenly city at the pole.[62] The Romans, like the Etruscans, prayed facing north, the seat of the gods. In their rites of augury, they divided the sky into four parts, and signs appearing in the north quadrant were the most favorable.[63] Servius called the north "Domicilium Jovis" (House of Jove).[64]

The nineteenth-century scholars who explained the importance of the north to early humanity have been sadly neglected. Isaac N. Vail wrote:

Can we fail to see here the north stretched out over the empty space? That space will have to be conceded, and conceded it becomes the foremost, the chief, the first, the arch, and the beginning of the heaven, that grew and grew until it filled all space.
(*The Waters above the Firmament* 1874)

Dr. William F. Warren, a former president of Boston University, wrote and lectured on the subject of humanity's north polar origin.

The religions of all ancient nations ... with a marvelous unanimity associate the abode of the Supreme God with the North Pole, the center of heaven, or with the celestial space immediately surrounding it.
(*Paradise Found* 1885)

Author John O'Neill has suggested that the solar worship of Egypt was a late substitute for their original cosmic religion:

> The Most High, the deity symbolically worshipped on High Places, was the god of the Polestar, who was seated at the highest celestial spot of the Cosmos, the North Pole of the heavens. (John O'Neill, *The Night of the Gods,* Vol. I 1893)

The Mystery of Zion

Out of Zion, the perfection of beauty, God hath shined.
—Psalm 50:2 (KJV)

Towards the North, life shall be planted in the holy place, the habitation of the everlasting King.
—Enoch 24:9b

The Hebrew name of the Throne of God is stated in Psalm 132:13, 14. "For the Lord has chosen Zion; He has desired it for his dwelling, saying, 'This is my resting place forever and ever; here will I sit enthroned for I have desired it.'" Psalm 99:2 proclaims, "The Lord is great in Zion; and He is high above all the peoples" (NKJV). Zion, "the perfection of beauty," is in the north. "Beautiful in elevation, the joy of the whole earth, is Mount Zion on the sides of the *north,* the city of the great King" (Psalm 48:2). God created the world from Zion and will renew it out of Zion. "Your watchmen shall lift up their voices, with their voices they shall sing together; for they shall see eye to eye, when the Lord brings back Zion" (Isaiah 52:8 NKJV). The salvation of Israel will come from Zion (Psalm 53:6).

What does the word Zion mean? Zion צִיּוֹן, from the primitive root *tsîyr,* means "a peg, a nail, a hook or a hinge." This root corresponds to the Greek *ouras* (socket, pivot), which is related to Ouranos (heaven). Ouras is similar in meaning to the Hebrew letter *vav* ו (nail). Each letter in Hebrew is also a word. One letter, *peh* פ, makes Zion the word for north: Zaphon צָפוֹן. Zaphon is the north wind. Mount Zaphon is the Phoenician Mount Zion. For these reasons (and others explored later), it is logical to conclude that the original meaning of Zion must be "Nail of the North."

The Nail in His Holy Place

And now for a little space grace hath been showed from the Lord our God, to leave us a remnant to escape and to give us a *nail in his holy place,* that our God may lighten our eyes, and give us a little reviving in our bondage.
—Ezra 9:8 (KJV)

A nail was a prime symbol of the supreme God all over the ancient world. Nails were highly prized and worn as symbols of blessing. Each ancient city had its own foundation nail (see Sumerian Foundation nail of the <u>E-ninnu</u>). Ethnographer and folklorist Uno Holmberg explains the celestial origin of these traditions:

> In the middle of the sky, or in the north, the heavens are affixed to a nail in such a manner that they are able to revolve around the nail, the revolving causing the movement of the stars. (Uno Holmberg, "Finno-Ugric Mythology" Vol. 4, *Mythology of All Races* 1916)

The Estonians and Finns speak of the "Nail of the Sky," and "Pohjanael" (Nail of the North). The Lapps honored the supreme God by erecting a wooden pillar with an iron nail fixed to the top and say that when Boahje-naste (North-nail, North Star) is shot down by Arcturus, the archer, the heavens fall, crushing the earth and setting fire to everything, ending a world age.[65]

According to the Agaria people of central India, the breaking of an iron nail caused the golden age town of Lohripur to be flooded.[66] The *Rg Veda* calls this nail "a linch-pin":

> As on a linch-pin, firm rest things immortal: he
> who hath known it let him here declare it.
> —*Rg Veda,* book 1 Hymn 35
> (Ralph TH Griffith, trans. *Hymns of the Rg Veda* 1889)

Whirling dervishes are Sufi ascetics who practice a very ancient, mystical art. They begin the dance bowing with crossed arms, then whirl to the right on one spot. Their costumes are often very colorful, but some wear long, white, flaring robes. Originally there were twelve orders; the oldest wear the Kulah, a high conical felt hat pointing to the zenith. The circular dance spinning against the sun involves thirty revolutions per minute, thus imitating the Star of God turning on its axis while remaining at rest at the Arx of the heavens. The dervish wears a *cherkha* (wheel).[67]

The Etruscans gathered annually in the center of their confederation for a sacred ceremony. One member chosen to act as head slowly drove the great wooden "Year Nail" into the wall of the inner sanctuary of the temple.[68] The words *nail* and *key* are the same in Latin—*clavis*. Clavis and the Greek *cataclysm* derive from an earlier root, *kleu*, (hook or peg).[69] The Hebrew letter *vav* ו (nail) originally looked like this: ן.[70] It is the third letter in the Holy name יהוה YHWH. *Vav* (or *Waw*) as the connecting link or uniter[71] (usually translated "and") naturally has high symbolic value.

Isaiah speaks of the removal of the nail. "In that day, says the Lord of hosts, the *nail or peg that was fastened into the sure place* shall give way and be moved and be hewn down, and fall, and the burden that was upon it shall be cut off; for the Lord has spoken it" (Isaiah 22:25 AMP). The burden upon the nail was this world, and our northern sky is now "an empty place" (Job 26:7). For centuries, Jewish pilgrims have mourned the destruction of the temple at the Wailing Wall. Believing the divine presence is revealed in its north corner, they place nails in crevices to fulfill Isaiah 22:25.[72]

This prophecy also speaks of Christ, who was "cut off from the land of the living." He is the nail upon which everything hangs and turns. "And I will fasten him as a *nail in a sure place*, and he shall be for a glorious throne to his father's house. And they shall hang upon him all the glory of his father's house" (Isaiah 22:23, 24a KJV). The prophet Zechariah uses similar terms. "From Judah will come the cornerstone, from him the *tent peg,* from him the battle bow, from him every ruler" (Zechariah 10:4).

The Lamb of God stands in the center as the mediator between God and humanity, nailed to a cross between two sinners. His chosen place is the heart of the person and the sky. "Then I (John) turned to see the voice that was speaking with me, and when I turned I saw seven golden menorahs, and in the *middle* of the menorahs I saw someone like the Son of Man, clothed to His feet and girded at the chest with a golden belt … and His face was as the sun shines in its power *at noon*" (Revelation 1:12, 13, 16 ONMB).

Chapter 4

The Golden Age of God

The Four Ages

The words of the Lord are flawless, like silver purified
in a crucible, like gold refined seven times.
—Psalm 12:6

According to our ancient forebears, the history of humanity is one of regressive degradation. They divided it into four cosmic ages, each one associated with a metal of successively decreasing value: gold, silver, bronze, and iron. The closer the metal is to God, the more precious it is. Gold is the metal of the Holy of Holies. The tradition lives on in the Olympic games—the best athlete gets the gold, the second best gets the silver, and the third gets the bronze. Iron, used in weapons of war, was of less value.

In the book of Daniel, metals declining in value represent successive empires. The Lord revealed to him that Nebuchadnezzar's dream of a statue of gold, silver, bronze, and iron is a prophetic view of the future. The Babylonian Empire is the golden head, the Persian Empire is the chest of silver, and the bronze belly is the empire of Alexander the Great. The iron legs are the east and west divisions of the Roman Empire. The toes, a mixture of iron and clay, represent the international disunity at the end of the Iron Age (Daniel 2:31–43).

Egypt: First Time

> I will refine them like silver and test them like gold. They will
> call on my name and I will answer them. I will say, "They are
> my people," and they will say, "The Lord is our God."
> —Zechariah 13:9b

By all accounts, the best age was the first, the idyllic time before the Flood. All nations and tribes described it similarly. The Egyptian First Time was Tep Zepi (from *tep*, meaning head, beginning, best). The New Year festival Tep renp-t, like Rosh Hashanah, is the head of the year.[1] In First Time, Ra, from *ara* (create), formed a universe different from the present world, which he governed from the prince's palace.[2] This was the ideal to which all others were compared:

> The reign of the god Ra, who inaugurated the existence of human
> life, was a golden age to which they continually looked back
> with regret and envy ... Its like has never been seen since.
> —François Lenormant, *The Beginnings of History*

Ra was the shepherd of his flock. The people had a childlike nature and enjoyed a continual state of wellbeing, which lasted until the ingratitude of humanity caused Ra to withdraw himself beyond reach. In various regions, it was the Time of Horus, the Time of Ptah, the Time of Osiris, and so on. All that was good and perfect was established in First Time.[3] The purpose of all ritual was its restoration. Its events were regularly celebrated and reenacted.[4]

Rome: Golden Time

> He raises the poor from the dust and lifts the beggar from the ash heap,
> to set them among princes and makes them inherit the throne of glory.
> —1 Samuel 2:8 (NKJV)

The first-century BC Roman poets Horace and Virgil called it Tempus Aureum (Golden Time), and Saturnia Regna (Reign of Saturn) respectively:

Before Jupiter no farmers plowed the fields; it was not even right to set up property markers or to divide a field with a boundary line; men used to seek for the common good and earth herself, uncommanded, bore all things only too generously. (Virgil, *Georgic* 1)

Ovid (43 BC–?) described it in detail:

Golden was the first age, which with no one to compel, without a law, of its own will, kept faith and did the right. There was no fear of punishment, no threatening words were to be read on brazen tablets; no suppliant throng gazed fearfully upon its judge's face; but without judges lived securely ... There was no need at all of armed men, for nations secure from war's alarms, passed the years in gentle ease. The earth herself without compulsion, untouched with hoe or plowshare, of herself gave all things needful. And men content with food which came with no one's seeking, gathered the arbute fruit, strawberries from the mountainside cornel-cherries, berries hanging thick upon the prickly bramble, and acorns falling from the spreading tree of Jove. Then spring was everlasting, and gentle zephyrs with warm breath played with the flowers that sprang unplanted. (Ovid, *The Metamorphosis* trans. FJ Miller)

The original site of Rome was Saturnia. Ovid called it second only to heaven itself. Another name for it was Latium (hidden).[5] The Capitoline Hill was the Mount of Saturn.[6] According to Macrobius (AD fourth century), the Temple of Saturn was the public treasury, for in the time of Saturn, there were no thefts, and there was no slavery. Saturn's reign ended when he disappeared.[7] Frazer mentions the high places erected in his honor:

At last the good king, the kindly king, vanished suddenly; but his memory was cherished to distant ages, shrines were reared in his honor, and many hills and high places in Italy bore his name.[8]

During Rome's annual New Year feast, the Saturnalia, slaves enjoyed complete equality.[9] The king was lifted up to make white crosses with chalk on the ceiling, and bonfires were lit on every hill.[10] It was an offense to begin a war at this time or to punish criminals.[11]

Greece: The Age of Kronos

A thousand years in your sight are like a day that has
just gone by or like a watch in the night.
—Psalm 90:4

The Greeks knew four ages beginning with the Age of Kronos.[12] Kronos, from *krounos* (source, spring), the Father of all gods and human beings, was "enthroned at the North Pole."[13] Greek and Roman historians identified El as the Creator and father of humanity and identified him with Kronos, Saturn, Ra, and Anu.[14] In his Greek history, the Babylonian priest Berosus (third century BC) identified Kronos as the god who warned Xisuthros of the Flood and told him to build a vessel for his family and all the animals and birds.[15] The Kronalia feast celebrated the age when master and servant shared a common meal.[16] The Greek poets praised its ideal conditions:

> First of all the deathless gods who dwell on Olympus made a golden race of mortal men who lived in the time of Cronos when he was reigning in heaven. And they lived like gods without sorrow of heart, remote and free from toil and grief: miserable age rested not on them; but with legs and arms never failing they made merry with feasting beyond the reach of all evils. When they died, it was as though they were overcome with sleep, and they had all good things; for the fruitful earth unforced bare them fruit abundantly and without stint. They dwelt in ease and peace upon their lands with many good things, rich in flocks and loved by the blessed gods. (Hesiod, *The Works and Days* trans. Hugh G. Evelyn-White)

In his *Laws*, Plato praised the former happy life of humankind:

> In the primeval world, and a long while before the cities came into being, whose settlements we have described, there is said to have been in the time of Cronos a blessed rule and life, of which the best-ordered of existing states is a copy. ("Laws" trans. Benjamin Jowett)

And in *The Statesman*:

> And there was no violence among them, or war, or devouring of one another. Their life was spontaneous, because in those days God ruled over man; and he was to man what man is now to the animals. Under his government there were no estates, or private possessions, or families; but the earth produced a sufficiency of all things, and men were born out of the earth, having no traditions of the past; and as the temperature of the seasons was mild, they took no thought for raiment, and had no beds, but lived and dwelt in the open air. Such was the age of Cronos, and the age of Zeus is our own.
> ("The Statesman" trans. Benjamin Jowett)

Persia: The Reign of Yima

> Jesus said, "I am the good shepherd. The good
> shepherd lays down his life for the sheep."
> —John 10:11

Persian cosmic history is a battleground between good and evil divided into four eras of three thousand years each. The first two were concerned with creation. In the third, "The wills of the uncreated Creator and Ahriman (the devil) are mixed in the world."[17] It is the fourth era that is subdivided into four ages associated with metals: gold, silver, steel, and iron. Each was a step downward materially and morally. The last age is the time when evil will assault the world with renewed strength.

The Zend Avesta speaks of Yima the Splendid (Old Persian Yima Xsaeta). He sat on a throne as bright as the sun at the pole of the heavens. He was the good shepherd, the father of the human race. Peace and plenty characterized his thousand-year reign. Demons, untruth, hunger, sickness, and death were unknown. There was "neither cold nor heat, old age nor death, nor envy."[18] The sole possessor of the sun's eye, Yima had the power to make human beings and animals immortal and keep the waters and plants from drying up.[19] His New Year feast, Nauroz, is a time for gift giving. The prototype of all kings, his reign ended in a great cataclysm.[20]

India: The Krita Yuga

I see him, but not now; I behold him, but not near—a star will
come out of Jacob, and a scepter will rise out of Israel.
—Numbers 24:17

The sacred texts of India tell of four *yugas* (ages) with decreasing lengths of years that end in a Mahapralaya, a total dissolution of the cosmos. The first and best was the Krita-yuga, from *kri* (to make). It is also called Satya (real, true). Only one god was worshiped, and *dharma* (righteousness) walked on four legs. Human beings were healthy, virtuous, and prosperous,[21]

> In the Satya yuga, the people were very happy. There were no inferiors or superiors. Everyone was equal. The climate was not hot or cold. There was no hatred or jealousy. Hunger and thirst were unknown. The earth yielded a vast supply of juices and mankind enjoyed living on this. There was no need to build houses, because no one needed protection from the elements or each other. People lived on the beaches. There was no concept of sin (papa) and merit (punya). (*The Linga Purana*)

America: The Blessing Way

Your love, O Lord reaches to the heavens, your faithfulness to the skies.
—Psalm 36:5

The Navajo or Diné people of the American Southwest describe a series of four worlds. The first was "the Age of Beginnings," or "the Blessing Way," which ended in a great flood.[22] The Hopi describe four declining worlds symbolized by gold, silver, copper, and mixed metals. The first world, Tokpela, was yellow, and its mineral was gold. The people of this time were pure and happy, living in harmony with the animals. They enjoyed an abundant water supply and plentiful corn.[23] Similarly, the Indians of the central Andes speak of world ages ending in cataclysms *(pachakuti)*, which punished those ruling the land up to that time.[24]

The Aztecs called their ages "suns." Each was assigned to one of the four directions and elements: earth, wind, fire, and water, and ends with its extinction. At the close of the last sun, Nahui-atl (four water), the sky flooded the earth. One man, Tezpi, and his family and animals escaped after being warned to build a vessel out of a hollowed-out cypress. After the rain stopped, Tezpi sent out a vulture, which failed to return on account of the many floating corpses. He then sent out a hummingbird, which returned with a leafy branch.[25]

Australia: The Dreamtime

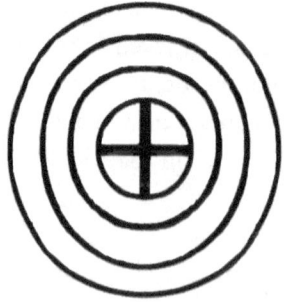

Aboriginal symbol of the ancestral land (ES)

The dreamtime was the first age when the ancestral beings lived. The dreaming tree of life, which grew in the sky world, was in place, and the morning star rested upon it. The earth was rich with plentiful water, and people enjoyed abundant food and lived without conflict. The righteous, spiritual energy of the dreamtime is called *djang*.[26] At the close of this epoch, the Sky Beings disappeared, and the physical landscape changed. The dreamtime symbolizes the foundation of all tribal life, and the religious word *dreaming* is termed "the Law." It becomes accessible in the present only through ritual.[27]

The Peace of the Golden Age

And a little child will lead them. The cow will feed with the bear, their young will lie down together, and the lion will eat straw like the ox.
—Isaiah 11:6, 7

During the golden age, nourishment was provided without bloodshed. Vegetarianism was God's original intention for both humans and animals: "Every tree that has fruit with seed in it. They will be yours for food. And to all the beasts of the earth and all the birds of the air and all the creatures that move on the ground—everything that has the breath of life in it I give every green plant for food" (Genesis 1:29, 30). Since there was no predation, there was no need for camouflage, and birds and beasts were brightly colored.

There was perfect communication between humans and animals. Eve expressed no surprise when a serpent spoke to her, and Noah kept the peace among the animals on the ark. The Chinese sage Kwang-tze (fourth century BC) spoke of this peace:

> The men of old, while the chaotic condition was yet undeveloped, shared the placid tranquility which belonged to the whole world ... not a single thing received any injury, and no living being came to a premature end. (*The Texts of Taoism* Volume I trans. James Legge)

During the happy age, animals were well treated, say the people of eastern Tibet. Afterward, the layer of blessings protecting the earth was eaten away by the grazing stars. Then the beasts were scorned and mistreated because they lost the power of speech.[28] The Sumerian hymns agree. "There was no fear, no terror, man had no rival."[29] According to the Caribs of Surinam,

> In a time long past, so long past that even the grandmothers of our grandmothers were not yet born, the world was quite other than what it is today: the trees were forever in fruit; the animals lived in perfect harmony, and the little agouti played fearlessly with the beard of the jaguar; the serpents had no venom; the rivers flowed evenly without drought or flood. (HB Alexander, "Latin American Mythology," Vol. 11, *Mythology of All Races*)

The Cheyenne tell of huge tame animals—a milk cow four times larger than now—all "destroyed in a great flood."[30]

In Southeast Asian tradition, rice grew by itself, and the people could help themselves to as much as they needed. One grain was said to be as big as a bowl, enough for an entire family. A ball of rice the size of a coconut was memorialized in a special pagoda.[31]

In the Americas, we find similar traditions of abundant food. The Aztecs told Franciscan Bernardo de Sahagun,

> In the days of Quetzalcoatl there was abundance of everything necessary for subsistence. The maize was plentiful, the calabashes were as thick as one's arm, and cotton grew in all colors without having to be dyed. A variety of birds of rich plumage filled the air with their songs, and gold, silver, and precious stones were abundant. In the reign of Quetzalcoatl there was peace and plenty for all men. But this blissful state was too fortunate, too happy to endure. (Louis Spence, *The Myths of Mexico and Peru* 1913)

The Roman poet Virgil recalls when wool came in color,

> But in the meadow shall the ram himself,
> Now with soft flush of purple, now with tint
> Of yellow saffron, teach his fleece to shine.
> While clothed in natural scarlet graze the lambs.
> (*Eclogue* 4)

The Mist World

I came forth from the mouth of the Most High, and covered the earth
like a mist. I dwelt in high places, and my throne was in a pillar of cloud.
—Ecclesiastes 24:3–4

In Genesis 1:31, God looked at everything He made and declared it *"tov
me'od"* (very good), but ever since the Flood, terrible storms have ravaged the
earth, most of which is uninhabitable. Obviously, the earth's present climate
is not what God originally intended. The quick-frozen fossils of tropical flora
and diverse fauna in the Arctic bear witness to a former temperate climate.
The Papago of Arizona say:

> The Great Spirit made the earth and all living creatures
> before he made man ... Those first days of the world were
> happy and peaceful. The sun was then nearer the earth
> than he is now: his rays made all equable and clothing
> superfluous. Men and animals talked together: a common
> language united them in bonds of brotherhood. But a terrible
> catastrophe put an end to those golden days. A great flood
> destroyed all flesh wherein was the breath of life. (James G.
> Frazer, *Folklore in the Old Testament*)

In Tibetan tradition, there were no storms, cold winters, or hot summers.
No rain or snow fell, and tea grew on the hills by itself. No one had anything to
fear.[32] What unique atmospheric conditions generated this perpetual spring?
All accounts describe abundant water but no precipitation. Wherever we look,
we find the primeval world watered by a fine mist cloud. As stated in Genesis,

> For God had not caused it to rain upon the earth: and there
> was not a man to till the ground; but there went up a mist
> from the earth, and watered the whole face of the ground.
> (Genesis 2:5, 6 KJV)

In *The Odyssey*, Homer tells of a haze that hung over the land. "No
snow is there nor yet any great storm nor any rain" (4. 561–9). In Norse
tradition, those living before Ragnarök were the Nibelungen (children of
the mist-world). The Egyptians called all moisture "the effusion of Osirus,"

and a water jar honored him in processions.[33] Four Canopic jars were used in Egyptian burial rites. Australian tribes say the dreamtime was dressed in "a curtain of rain."[34] The Kojiki of Japan speaks of a "heavenly mist deity" Ame-No-Sa-Giri-No-Kami.[35]

The Zuni of New Mexico refer to "the Place of Mist on the Waters" when fog *(shipolo)* rose like steam.[36] The Great Spirit Awonawilona created life and light within himself and produced a "fog-shield" of rising and falling mists and streams that promoted growth.[37] The Irish speak of Magh Da Cheo (the Plain of the Two Mists).[38] The Aztecs describe living in *ayauhcalli* (house of mist), where happiness was purer, less changeable. "Here people played leap frog, chased butterflies and sang songs."[39] The Inuit of Alaska say:

> In the days of the first people on the earth plain, there were
> no mountains far or near. No rain ever fell and there were
> no winds. The sun shone always very brightly.
> (Katherine Berry Judson, Ed. *Myths and Legends of Alaska*)

The antediluvian climate was of an entirely different order and magnitude. The divine designer created a cloud canopy over the preflood earth, the most efficient irrigation and cooling system ever devised, an interplanetary greenhouse. In Isaac Newton Vail's *The Waters above the Firmament,* he discusses the "Chronian Canopy" and suggests that increased water vapor and CO_2 would recapture radiated heat, creating a greenhouse effect resulting in a temperature equilibrium.[40] Storms form as a result of temperature differences.

The cloud canopy was a world heater by day and a world cooler by night. As the temperature fell, moisture absorbed by the air became a mist on the ground. The Persians describe aerial waters that promoted growth,

> As the sea Vourukasha is the gathering place of waters, rise
> up, go up the aerial way and go down on the earth; go down
> on the earth and go up the aerial way. Rise up and roll
> along, thou in whose rising and growing Ahura Mazda made
> everything grow. (Fargard 21. Waters and Light)

Among all tribes, fragrant smoke from burning incense symbolized these rising and flowing mists (Leviticus 24:7). The canopy shimmered like gold during the day. Vail suggested this was the reason for the name of the age itself.[41]

In *The Aeneid*, Virgil wrote of "the Lord of Heaven's Almighty Throne speaking as from a golden cloud."[42] Unlike any other planet in the known universe, 71 percent of the earth's surface is under water.

Without the canopy, we cannot account for the worldwide flood affirmed in the traditions of every nation. Jesus spoke of a global flood (Luke 17:26). There is not enough water in our present atmosphere for a rainfall sufficient to cover the entire earth (Genesis 7:11, 12). At the chosen time, the canopy waters were condensed and released, drowning all life and totally altering its climate. Remnants of the canopy remain in the upper stratosphere as water vapor ions.[43] The Lord will restore His canopy, "I will be still and I will look on them from my dwelling place like clear and glowing heat in sunshine, like a fine mist-cloud in the heat of harvest" (Isaiah 18:4 AMP).

Then the Lord will create over all of Mount Zion and over those who assemble there a cloud and smoke by day and a glow of flaming fire by night; over all the glory will be a *canopy*. It will be a shelter and shade from the heat of the day, and a refuge and hiding place from the storm and rain (Isaiah 4:5, 6). This will require the ocean waters to be evaporated and returned to the atmosphere. Global volcanic activity will bring this about (see 2 Peter 3:10.

How to Live a Thousand Years

Chinese sign of longevity, Han Dynasty tile
"Myriad Years without Limit." (ES)

But, beloved, be not ignorant of this one thing, that one day is with
the Lord as a thousand years, and a thousand years as one day.
—2 Peter 3:8

God's plan for humanity originally was eternal life, but after the fall, it was a lifespan of one thousand years, one of His days. The oldest man in the Bible was Noah's grandfather. "And all the days of Methuselah were 969 years: and he died" (Genesis 5:27). Ancient historians Hesiod, Aerates, Ovid, Virgil, Claudius, and others described the men of their time as "short-lived" compared to their ancestors. Their senses and procreative powers lasted for hundreds of years.[44] Plato wrote:

> Moreover, the temperament of their seasons is such that they have no disease, and live much longer than we do, and have sight and hearing and smell, and all the other senses, in far greater perfection, in the same proportion that air is purer than water or the ether than air. ("Phaedo" trans. Benjamin Jowett)

In Teutonic lore, people lived much longer in the golden age. They were innocent and dwelt in peace.[45] In Buddhist tradition, the progressive decadence of humanity is marked by a constant lessening of human life.[46] The Chinese sages, during this age, lacked the three evils:

> The sage, after living a thousand years, ascends among the Immortals and arrives at the place of god, free from the three evils of disease, old age, and death. (*The Texts of Taoism* Volume I trans. James Legge)

The Hopi assert that the Creator planned to keep their ancestors free from disease and give them long life spans.[47] The Maidu of California hold similar traditions. They said that all fruits were easily obtained and no one ever got sick.[48] In Old Persian tradition, "Fathers and sons alike had the air of young men of fifteen, as long as Yima reigned."[49] The Kalmucks, a Mongol tribe, shared this same tradition:

> Another consequence of the fall, according to the Kalmucks, was the shortening of the age of man and a reduction in his size. (Uno Holmberg, "Siberian Mythology" Vol. 4 *Mythology of All Races* 1916)

According to the *Mahabharata,* a lifespan of one thousand years was normal during the Krita Age.[50] Men were free of disease and fatigue:

> They suffered no impediments, no susceptibilities to the pairs of opposites (like pleasure and pain, cold and heat) and no fatigue. They frequented the mountains and seas, and did not dwell in houses. They never sorrowed, were full of the quality of goodness, and supremely happy; they moved about at will and lived in continual delight ... Produced from the essence of the earth, the things, which those people desired sprang up from the earth everywhere and always, when thought of. That perfection of theirs both produced strength and beauty and annihilated disease. With bodies, which needed no decoration, they enjoyed perpetual youth. (*Vaya Purana*)

In AD 93, Josephus wrote of the extended lives of the patriarchs and names other historians who so testify but whose works are no longer extant. He mentions superior food,

> But let no one upon comparing the lives of the ancients with our lives, and with the few years which we now live, think that what we have said of them is false; or make the shortness of our lives at present an argument, that neither did they attain to so long a duration of life; for those ancients were beloved of God, and made by God himself; and because their food was then fitter for the prolongation of life, might well live so great a number of years ... for even Manetho, who wrote the Egyptian history, and Berosus, who collected the Chaldean monuments, and Mochus and Hestiaeus, and besides these, Hieronymus the Egyptian, and those that composed the Phoenician History, agree to what I here say: Hesiod also, and Hecataeus, and Hellanicus and Acusilaus; and besides these, Ephorus and Nicolaus relate that the ancients lived a thousand years. (*Antiquities*, book 1, chapter 3)

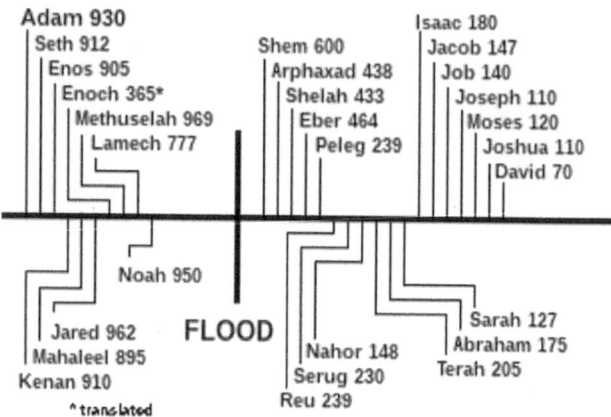

Timeline of the patriarchs
Author: equal illustrations

Many diseases have environmental causes, and beneficial plants supplying vital nutrients may now be extinct. There was a continuous decline in longevity in the lives of the patriarchs. Post-flood, the amount of C-14 in the atmosphere would increase in each generation, producing a corresponding reduction in longevity. Cosmic rays produce mutations in genetic material, the vast majority of them harmful. With the collapse of the canopy, a much greater amount of cosmic radiation penetrated the atmosphere, causing mass extinctions of species.[51]

Ovid wrote of the appearance of new life forms:

> When therefore the earth covered with mud from the recent flood, became heated up by the hot and genial rays of the sun, she brought forth innumerable forms of life, in part of ancient shapes, and in part creatures new and strange. (Ovid, Metamorphoses trans. FJ Miller)

Greek historian Diodorus Siculus, who wrote histories between 60 and 30 BC, spoke of mutations after the Flood.

> Some maintain the destruction of living things was complete and the earth then brought forth again new forms of animals. (*Library of World History*, volume 1)

The Greek tragedian Euripides wrote,

> Many are the shapes of heaven's denizens, and many a thing they bring to pass contrary to our expectation; that which we thought would be is not accomplished, while for the unexpected God finds out a way. (*Andromache*)

In *The Waters Above,* Joseph Dillow, PhD, observed that under the canopy, no carbon 14 would be formed because all the nitrogen would have been shielded from the effects of cosmic radiation. In the manufacture of proteins and DNA, the body uses C-14, but it spontaneously decays into N-14, emitting ionizing radiation. Due to the temperature inversion resulting from the absorption of terrestrial radiation, there would have been little diffusion of the lower atmosphere into the upper layers. Therefore, preflood samples dated by C-14 would appear to have infinite age.

If the atmospheric radiocarbon at an animal's death was lower than today, it would appear that much more of the animal had decayed, making it seem much older than it is. A sample dated at thirty-five thousand years would be only 5,251 years old.[52] A quick-frozen Siberian mammoth hide was dated at about forty thousand years, but the buttercups and other non-arctic plants in his stomach protected from the atmosphere dated to about six thousand years.[53]

In their book *The Genesis Flood,* John C. Whitcomb and Henry M. Morris explain why radiocarbon dating is dependent on certain inherent assumptions, all of which are highly questionable during the applicable period:

1. That the C-14 concentration in the carbon dioxide cycle is constant.
2. That the flow of cosmic rays has been constant for centuries (and dead organic matter is not later altered with respect to its carbon content by any biologic or other activity).
3. That carbon dioxide in the oceans and atmosphere has been constant with time.
4. That the immense reservoir of oceanic carbon has not changed in size.
5. That the rate of formation and rate of decay of radiocarbon atoms have been in equilibrium.[54]

The Real Reason Noah Got Drunk

Noah, a man of the soil, proceeded to plant a vineyard. When he drank
some of its wine, he became drunk and lay uncovered in his tent.
—Genesis 9:20, 21

The Aztecs said that Ehecatl (Quetzalcoatl) put himself in harm's way by
bringing down to earth intoxicating liquor made from the maguey plant,
so that men would have some relief from their sweat and toil.[55] According
to the Greek poet Nonnus (c. AD fourth century), wine was given as divine
assistance after the Flood to comfort the people in their sorrow and loss.[56] In
his Symposium, Plato said that wine was unknown in Kronos's time (2, 4).
The legendary Siberian leader Kezer-Tshingis-Kaira-Khan was said to have
invented liquor:

> After the Flood, *Kezer-Tshingis-Kaira-Khan* recreated
> everything we now see around us. He is especially mentioned
> as having taught people how to prepare strong drinks. (Uno
> Holmberg, *Finno-Ugric Mythology* Vol. 4, *Mythology of All
> Races* 1916)

Before the Flood, the rate of fermentation was so slow it was barely
noticeable, one of the reasons no fermented foods are permitted on the
High Holidays, the Feast of Unleavened Bread, and Passover, except for
wine. The higher surface atmospheric pressure before the Flood resulted in
an increased partial pressure of CO_2. Higher atmospheric CO_2 makes plants
thrive, producing higher O_2 greatly slowing the fermentation rate. With the
condensation of the canopy, atmospheric pressure was suddenly reduced
from 2.18 atmospheres to its present level of one. At lower pressures, the
rate of alcohol production would greatly increase. Noah knew nothing of
this and became drunk after harvesting his first vineyard.[57]

When There Were Giants on Earth

There were giants in the earth in those days and also after that ...
—Genesis 6:4a KJV

In rabbinic tradition, human stature was shortened after the Flood.[58] This seems to be a universal tenet. In his "Laws," Plato records that the men of his day were inferior to those that lived before the deluge. The Sumerian texts agree that the remaining offspring of all living things was "tiny," and they were not returned to their previous state.[59] St. Augustine wrote of this decline:

> For though the bodies of ordinary men were then larger than ours, the giants surpassed all in stature ... The younger Pliny, a most learned man, maintains that the older the world becomes, the smaller will be the bodies of men ... Homer in his poems often lamented the same decline. (*The City of God* Book 15)

According to the Hindu texts, "Men were tall and long-lived" during the Krita Yuga but in successive ages became "dwarfed and feeble."[60] The Babylonian priest Berosus makes similar statements:

> The men were of extraordinary stature and strength and as retaining in lessening degree these characteristics until some generations after the flood. (*Fragments Cosmogonique de Berose*)

The tribes of the Pacific Northwest say their ancestors were as "tall as the pine and fir trees" covering the hills.[61]

Dr. Dillow suggests that giants thrived in increased partial pressure of oxygen, enabling them to supply more of it to their larger tissue mass. Higher oxygen tension promotes healing and reverses the effects of aging.[62] Only a hyperbaric chamber can deliver oxygen at pressures greater than one atmosphere. Traditions abound of men becoming too heavy to use their former abilities. The Lao of Cambodia say eating a gourd made men too heavy to return to God's Land. The Kachins of Burma say their ancestors were sinless, but eating rice made them heavy and gross.[63] A tradition from Sudan asserts that people could once get to the moon by a road but became too heavy to use it.[63]

The stresses of increased gravity are a huge factor in aging. According to Virgil, an ancient strong man could pick up a heavy landmark stone as he fought and hurl it with ease,

> There he ceased, and, glancing round,
> Spied a vast stone, an ancient stone and vast,
> Which chanced to lie there, for a landmark pitched …
> Scarce could twelve of such as earth in bulk
> Now bears, upon their necks have heaved it; he
> Caught with quick hand and hurled it on the foe.
> (Virgil, *Aeneid* book 12)

After the Flood, the pull toward the center of the earth doubled. An object one man formerly could lift, many men now had to use great effort. The Inca priests told the Spaniards that the huge stone edifices around Lake Titicaca were built by a race that lived on earth before the present sun was in the sky. These giants could lift two-hundred-ton blocks of limestone with ease. On a hill above the city of Cuzco is <u>Sacsahuamán</u>, an immense fortified wall six hundred yards long, built in three lines of enormous stones, some twenty-seven feet in length. At Ollantay-tampu, forty-five miles north of Cuzco, is another of these fortresses with walls of stones weighing many tons perfectly cut and fitted together without mortar, and sites aligned astronomically.[64]

These constructions cannot be duplicated today even with the most advanced technology. In their essay "A Comprehensive Theory on Aging, Gigantism, and Longevity," Donald and Phillip Patten propose that intercontinental gigantism has little to do with genetics and everything to do with climate. Global decline in body size was sudden, recent, and concurrent with the decline in longevity. Both occurred with two parallel developments (estimated to be from 5,000 to 6,500 years ago), a rapidly changing climate, and the worldwide explosion of human population and culture.[65]

Age of Virtue

> He will judge the world in righteousness; he
> will govern the peoples with justice.
> —Psalm 9:8

In the records of all nations, human beings were moral at first. Roman orator and historian Tacitus (c. AD 55–117) wrote:

Mankind in the earliest age lived for a time without a single vicious impulse, without shame or guilt, and, consequently, without punishment and restraints. Rewards were not needed when everything right was pursued on its own merits; and as men desired nothing against morality, they were debarred from nothing by fear. When however they began to throw off equality, and ambition and violence usurped the place of self-control and modesty, despotisms grew up and became perpetual among many nations. (Tacitus, *The Annals*, book 3)

Josephus tells of the good character of Seth, son of Adam.

Now this Seth ... became a virtuous man; and as he was himself of an excellent character, so did he leave children behind him who imitated his virtues. All these proved to be of good dispositions. They also inhabited the same country without dissensions, and in a happy condition, without any misfortunes falling upon them till they die. (Josephus, *Antiquities*, book 1, chapter 2)

The Mayans say the first people were grateful to their Maker.

They were good and handsome people, 'Perfectly they saw, perfectly they knew everything under the sky, wherever they looked. The moment they turned and looked around in the sky, on the earth, everything was seen without any obstruction. They didn't have to walk around before they could see what was under the sky, they just stayed where they were. And they thanked their Modeler and Maker. (*Popul Vuh* trans. Dennis Tedlock)

The Chinese sages declared that human beings first existed in a condition of happy innocence.[66] "The flesh was obedient to the spirit," and society lived according to the rules of perfect virtue. This is the time when the star of virtue, *DeXing* 德星, shone in the heavens (notice that both characters contain the sign of God, and the first locates it over four pillars). The people then "were able to control all their natural passions."[67] The sage Kwang-tze describes the era:

> They were upright and correct, without knowing that to be so was righteousness; they loved one another, without knowing that to do so was benevolence; they were honest and leal-hearted, without knowing that it was loyalty; they fulfilled their engagements, without knowing that to do so was Good Faith; in their movements they employed the services of one another ... All creatures lived in companies, and their places of settlement were made near to one another. Birds and beasts multiplied to flocks and herds; the grass and trees grew luxuriant and long. The birds and beasts might be led about without feeling the constraint; the nest of the magpie might be climbed to, and peeped into. Yes, in the age of Perfect Virtue, men lived in common with the birds and beasts, and were on terms of equality with all creatures, as forming one family.[68]

Age of Vice

As surely as I live, declares the Sovereign Lord, I take no pleasure in the death of the wicked, but rather that they turn from their ways and live ...
—Ezekiel 33:11a

According to the Talmud, the antediluvians gradually forgot the source of their blessings. Under these ideal conditions, they worked so little that they took their benefits for granted and grew arrogant toward God.[69] Josephus wrote that Seth's descendants were righteous for seven generations, but over time, they began to dishonor God and showed little concern for justice toward other human beings.[70] The Book of Third Maccabees comments on their fate:

> You destroyed those who in the past committed injustice, among whom were even giants who trusted in their strength and boldness, whom you destroyed by bringing on them a boundless flood. (3 Maccabees 2:4)

The Hindu Puranas assert that the disorder of nature echoed a "spiritual degeneracy," a decrease in excellence and virtue among humans and the quality of life.[71] There was also a steady decline in Dharma (righteousness), and although by divine grace there is salvation for some individual sinners, this does not prevent the increase of social disorder.[72] Toward the end, a decline in virtue led to the pursuit of riches and honors, and the world descended into chaos.

The Chinese sages echoed this decline:

> Complacency (set) in, and hatred of one another produced mutual suspicions; the stupid and wise imposed on one another; the good and the bad condemned one another; the boastful and the sincere interchanged their recriminations; and the world fell into decay. Views as to what was greatly virtuous did not agree, and nature with its endowments became as if shriveled by fire or carried away by a flood. (*The Texts of Taoism* volume 1 trans. James Legge)

The Navajo say the first people were destroyed by a flood for committing adultery, practicing witchcraft, and defiling themselves with animals.[73] The Hopi say their ancestors followed the Creator's plan at first but gradually drew away from Him and divided into different races and tongues.[74] The Mayan *Popul Vuh* asserts that the first humans were like wood and forgot their Creator. Then, Heart of Sky destroyed them with a flood.[75] According to the Aztecs, "seeds of decay" brought the glorious reign of Quetzalcoatl to an end.[76] A poem overflows with regret.

> The plumes of the quetzal
> The works of iridescent jade
> All broken and gone
> The memory of a beautiful world
> god-filled, truth-filled …
> (Angel Maria Garibay Kintana, *Historia de la Literatura Nahuatl*)

In the Norse texts, before Ragnarök, incest and adulteries multiplied. Brothers fought each other with axes and swords,

Brothers shall strive | and slaughter each other;
Own sisters' children | shall sin together;
Ill days among men, | many a whoredom:
An axe-age, a sword-age, | shields shall be cloven;
A wind-age, a wolf-age, | ere the world totters.
Heaven yawned and was "rent in twain."
("The Poetic Edda" trans. Henry Adams Bellows)

An Invasion from Above

If God places no trust in his servants if he charges his angels with error,
how much more those who live in houses of clay, whose foundations
are in the dust, who are crushed more readily than a moth!
—Job 4:18–19

A Slavonic tradition tells of a rebellion in heaven:

When the Supreme God created heaven and earth one party
of the spirits who surrounded him revolted. He drove these
rebellious spirits from the sky and cast them to earth. (G.
Alexinsky, "Slavonic Mythology"[77])

Greatly accelerating humanity's downward plunge was an amazing occurrence. In Genesis 6:1, we read of rebellious angels who, endowed with free will like ourselves, invaded the earth. "When men began to increase in number on the earth and daughters were born to them, the sons of God saw that the daughters of men were beautiful, and they married any of them they chose." These were not human beings for the Hebrew term "sons of god" is used only for angels.[78] The Septuagint calls them both "angels of God" and "sons of God."

They are also called "angels" in the New Testament, "For if God did not spare the angels when they sinned but sent them to Tartarus, putting them into gloomy dungeons to be held for judgment; if he did not spare the ancient world when he brought the flood on its ungodly people, but protected Noah, a preacher of righteousness" (2 Peter 2:4, 5). "And the angels who did not keep their positions of authority but abandoned their own home-these he has kept in darkness, bound with everlasting chains for judgment on the great Day" (Jude 1:6).

The Dead Sea Scrolls calls them "Heavenly Watchers." They were giants as were their offspring. They had stubborn hearts and did not keep God's commandments. Therefore, their Maker's wrath was kindled against them, and all flesh on dry land perished.[79] The Egyptians also called them Watchers:

> Deliver me from the Watchers who bear slaughtering knives, and who have cruel fingers, and who slay those who are in the following of Osiris. May they never overcome me, may I never fall under their knives.
> (*The Egyptian Book of the Dead*)

These *neteru* (gods) were superior to humans but subservient to Ua Neter, the One Eternal God. The Egyptians described them:

> These had many forms and shapes, and could appear on earth as men, women, animals, birds, reptiles, trees, plants, etc. They were stronger and more intelligent than men, but they had passions like men; they were credited with possessing some divine powers or characteristics, and yet they could suffer sickness and die. (EAW Budge, *Osirus and the Egyptian Resurrection* Vol. 1)

In Genesis, those that rebelled are called Nephilim (fallen ones), from the Hebrew *nophel* (to fall). "The Nephilim were on the earth in those days—and also afterward—when the sons of god went to the daughters of humans and had children by them. They were the heroes of old, men of renown" (Genesis 6:4). The Midrash tells of their immorality with virgins, married women, men, and animals.[80] They taught their wives charms and enchantments and herbal remedies. Uzzah and Azazel taught magic arts, metalwork, jewelry, how to make and use weapons, and how to apply eye makeup. Shemhazai taught enchantments and root cutting, and Kokhviel taught men astrology. They consumed all the fruits of the efforts of men. They injured beasts, birds, fish, and reptiles and drank their blood. When ordinary men could no longer provide food for them and their offspring, the Nephilim turned against them and devoured them.[81]

Josephus writes of Noah's distrust of them,

Many angels of God accompanied with women begat sons
that proved unjust, and despisers of all that was good, on
account of the confidence they had in their own strength;
for the tradition is, that these men did what they did and,
being displeased at their conduct, persuaded them to change
their dispositions and their acts for the better; but, seeing
that they did not yield to him, but were slaves to their wicked
pleasures, he was afraid they would kill him with his wife
and children, and those they had married; so he departed
out of that land. (*Antiquities,* book 1, chapter 3)

The ancients called them "gods, immortals, titans." Greek, Roman,
Teutonic, and Hindu mythology is filled with tales of their lust for human
beings and the births of offspring who were "demi-gods." Hesiod referred to
"those deathless ones who lay with mortals and bare children like unto gods."[82]
The Polynesians said their first ancestors had children who "intermarried
with the gods," and these begat the nobility of the world.[83] The Australian
Aborigines tell of the Pungalunga men who were the cannibal giants of the
dreamtime. They hunted men for food and returned to their camp with
bodies tucked into their hair-string belts. They were destroyed after a great
battle.[84]

Among the ancient traditions of Anahuac (Mexico) were a special class of
spirits who descended to earth, but others say that purer spirits drove them out
of heaven. Their leader was Zoutemque.[85] They were envious of humans and
brought contagious diseases and injuries. They were described as a primeval
race of giants and fallen arrows from the sky. Their strength was so great that
they could pull up trees.[86]

Ixtlilxochitl has given us one of the most complete Mexican creation
accounts, receiving it directly from native sources. These stated that the
Toltecs credited Tloque Nahuaque (Lord of All Existence) with the creation of
the universe, including the stars, mountains, and animals. He made the first
man and woman, from whom all the inhabitants of the earth are descended.
This "first earth" was destroyed by the "water-sun." At the start of the next
epoch, the Toltecs appeared, and after many wanderings, they settled in
Huehue Tlapallan (Old Old Tlapallan). Then followed the "wind-sun,"
when mighty earthquakes shook the world and destroyed the earth-giants
(Quinames) analogous to the Greek Titans. They were a source of great
uneasiness to the Toltecs."[87]

The Navajo declare that the Yei spirits were kind to humans, but the Anaye were "alien gods" and "man-destroyers." Consorting with women, they resorted to evil practices, and their offspring were giants. The worst of them were slain by the "sons of the sun," but they weren't completely eradicated.[88] Demons arose from the dead bodies of the giants and remained a permanent evil force in the world causing "wickedness, disruption, and oppression."[89]

The Iroquois and Huron tribes describe the giants as powerful magicians and hunters. They had incredible strength, using trees as weapons with ease and stones as missiles. Because they practiced cannibalism, they were greatly feared.[90] The Pocomam, a Mayan tribe of the eastern Guatemala highlands, concur. Before the Flood, a cannibalistic humanity dominated the world. They had neither faith nor law.[91]

The tribes of the Pacific Northwest tell of demon-giants called Seatco who gave themselves up to wrongdoing. They could imitate the call of every bird and the sounds of wild beasts. If humans managed to kill one of them, they took savage revenge by killing twelve people. Some of them lived in the mountains above Spirit Lake. They came at night stealing all of the salmon the people had caught. They had a strong, unpleasant odor.[92]

Inca historian Cieza de León described the giants as having shoulder-length hair but no beards. Some dressed in animal skins, but others were naked. Many were involved in building projects. They expertly bored deep wells in rock to get water, which they lined with masonry. This water was cool and wholesome to drink. The giants ate so much that they exhausted the resources of the district. One giant could devour the meat of fifty ordinary human beings. They were also prodigious fishermen. The natives hated them because they killed their women in associating with them.[93]

The people banded together against the giants but could not overcome them. They were many years in the land. The natives avoided them because of their size. Because they lacked women and had no fear of God, they practiced immorality with each other.[94] The tribes of the central Andes say the giants were three times larger than humans, with huge appetites and horrible practices.

> They committed the sin of abomination, to which they gave themselves one to another with much bestiality, for they hated women to death, with whom they coupled only for reproduction, and if females were born, they suffocated them. So it was told, in the arms of the midwife. But these

abominations did not go without punishment, heaven being their executioner, which threw them down with lightning. (John Bierhorst, *The Mythology of South America*)

The Incas said that the giants had begun to fight each other.

At first all went well, but after a time the giants began to fight among themselves and refused to work. Viracocha decided that he must destroy them. (Gifford and Sibbick, *Warriors, Gods and Spirits from So. American Mythology*)

The Old English poem "Beowulf" (c. AD 700) tells of their judgment.

Hrothgar began to speak while he examined the ancient relic, upon which was inscribed an account of the origin of the primeval conflict; how when the swirling waters of the flood destroyed the giant race, they suffered sorely; because they were hostile to God, He paid them out with a deluge of waters. ("Beowulf" trans. David Wright)

Toward the end of the age, the world had become so contaminated and corrupt that only eight people were found worthy to be spared. "The Lord saw how great man's wickedness on the earth had become, and that every inclination of the thoughts of his heart was only evil continually" (Genesis 6:5, 6). "So God said to Noah, 'I am going to put an end to all people for the earth was full of violence'" (Genesis 6:13). An Egyptian funerary text in the tomb of Seti I explains why humanity was destroyed by a flood.

They have fought fights, they have upheld strifes, they have done evil, they have created hostilities, they have made slaughter, they have caused trouble and oppression ... [Therefore] I am going to blot out everything, which I have made. This earth shall enter in the watery abyss by means of a raging flood, and will become even as it was in primeval time. (*From Fetish to God in Ancient Egypt*[95])

The Nisqually Indians of the American Northwest say the giants ate all the fish and game and then began to eat each other. They had become so wicked that the great Changer Dokibatl sent a flood to exterminate them. All died except a woman and a dog.[96] The Nebraska Pawnee similarly testify:

> There was a time when Tirawa Atius had placed giant people on the earth. But the giants grew proud and had to be destroyed. Storms came from the Northwest, the waters rose and rain poured down. The race of giants was destroyed; the last of them died on a hill in Kansas. (Cottie Burland, *North American Indian Mythology*)

The Hopi say there were a few who still lived by the creation laws of Taowa. They were given an "inner vision" to follow a certain cloud during the day and a certain star at night. They were led to, and kept safe in, an "Ant kiva" while the rest of the world was destroyed in a terrible cataclysm in which all volcanoes in the world erupted.[97] Many skulls of these giants have been found and can be seen in museums in Peru. Their huge brain capacity is clearly evident in their vastly elongated craniums.

The Silver Age

> The Lord was grieved that he had made man on the
> earth, and his heart was filled with pain.
> —Genesis 6:6

Much less has been written down about the silver and bronze ages, so immense was the contrast after the golden. Hesiod writes:

> Next after these the dwellers in Olympus created a second generation of silver, far worse than the other. They were not like the golden ones either in shape or spirit. (Hesiod, *Works and Days* trans. Hugh G. Evelyn-White)

Now human beings needed shelter from the elements. Increased gravity and labor made rest essential. The people needed the help of animals to raise their food and carry their loads. Plato wrote in the *Statesman* of their initial helplessness when the food, which once grew spontaneously, failed, and they didn't know anything about agriculture and had never felt the pressure of necessity.

The Sumerians expressed their horror over the lower yield.

> The furrow's yield diminished, and forever after it was hard to extract. Even the control of heaven and earth was undone, the springs diminished, as the floodwater receded. I went back, and looked, and looked; it was very grievous. (Stephanie Dalley, *Myths of Mesopotamia: The Creation, The Flood, Gilgamesh and Others*)

Ovid laments all that was lost.

> After Saturn had been banished to the dark land of death, and the world was under the sway of Jove, the silver race came in, lower in the scale than gold, but of greater worth than yellow brass. Jove now shortened the bounds of the old-time spring, and through winter, summer, variable autumn, and brief spring completed the year in four seasons. Then first the parched air glared white with burning heat, and icicles hung down congealed by freezing winds. In that age men first sought the shelter of houses. Their homes had heretofore been caves, dense thickets, and branches bound together with bark. Then first the seeds of grain were planted in long furrows, and bullocks groaned beneath the heavy yoke. (Ovid, *The Metamorphosis,* book 1 trans. FJ Miller)

Meat was now needed to provide essential protein and minerals. The relationship between humans and animals changed.

> Then God blessed Noah and his sons, saying to them, "Be fruitful and increase in number and fill the earth. The fear and dread of you will fall on all the beasts of the earth, and on all the birds in the sky, on every creature that moves along the ground, and on all the fish in the sea; they are given into your hands. Everything that lives and moves about will be food for you. Just as I gave you the green plants, I now give you everything." (Genesis 9:1–3)

Fire was now needed for light, warmth, and cooking, and many tales of its origin arose at this time. According to the Altai Tatars,

> Mankind originally lived on vegetables and fruits and therefore neither needed fire nor missed it, but with the change in their manner of nourishing themselves fire became necessary for the preparing of food. It was then that Ülgen took two stones, a white one and a black one, and struck them together so that the spark, which flew from the sky to the earth set fire to the dry grass. From this man learned to strike fire. (Uno Holmberg, "Siberian Mythology" Vol. 4 *Mythology of All Races* 1916)

The Thunder-God

So radical were the changes in the sky and the environment that the survivors concluded that a powerful rain and storm god now reigned,

> An examination into the ancient annals of every people
> shows the remarkable fact that the name of a people's celestial
> deity always changed with the changes of skies.
> —Isaac N. Vail, *The Waters above the Firmament*

Although Moses uses both the names Elohim and YHWH, for He is the same god, some tribes concluded that a warlike son or brother drove out the benevolent God of the golden age. Plutarch said that Set, twin brother of Osiris, murdered him. Horus, son of Osiris, restores order after a long battle between them.[98] Plato said that Zeus, the thunderbolt hurler, succeeded his father Cronos,

> And is this cycle, of which you are speaking, the reign of Cronos, or our present state of existence?' No, Socrates, that blessed and spontaneous life belongs not to this, but to the previous state, in which God was the governor of the whole world ... ("Statesman" trans Benjamin Jowett)

The thunder-wielding rain god Jove drove the Roman Saturn from his sky-throne. At Ragnarök, Thor, the storm-bringer, took over heaven from the

Norse All-Father Odin. The stormy Indra removed Varuna from his celestial throne. The Persian Yima was cut in two by his brother. The Aztecs said that the warlike lightning hurler Tezcatlipoca drove the gentle Quetzalcoatl from the heavens.

The Hopi recall the Second World Topka as not as beautiful as the first world. Silver was its mineral, and blue was its color. The people now lived apart from the animals, who were no longer tame. They built homes and villages and trails between them. They stored food, began to trade and barter, and quarrels led to wars between them. The more things they obtained, the more they wanted. Most of them forgot to sing the joyful song of creation. Others laughed at the few who remembered to sing. Little by little, they all were drawing away from Taiowa and the rich spiritual life of the earlier age.[99]

The Hindu second age was the Treta marking the start of regression. Men quarreled and acted with ulterior motives.

> In the other (three), by reason of unjust gains (agama), Dharma is deprived successively of one foot, and through theft, falsehood, and fraud, the merit (gained by men) is diminished by one fourth (in each). (*Laws of Manu* trans. George Buhler)[100]

Humanity's lot was now hard work and suffering. Duty was no longer spontaneous but had to be learned. Life span and adherence to the holy Vedic doctrines is decreased by one quarter.[101]

The Bronze Age

> Even to your old age and gray hairs I am he, who will sustain you. I have made you and I will carry you; I will sustain you and I will rescue you.
> —Isaiah 46:4

During the Age of Bronze (also called brass or copper), warfare and impiety increased. The Hopi call this third world Kuskurza. Red was its color, and copper was its mineral. Large cities were constructed, entire civilizations, and it became more difficult to conform to the plan of Taiowa and sing his praises. The people engaged in sexual immorality and became more and more occupied with their own worldly plans.[102]

In the Hindu Vayu Purana, the third age, Dvapara lasted two thousand years. Throughout, vices and evils increased, and human life was further shortened.[103] Plato summarizes the decline:

> During a certain period God himself goes with the universe as guide in its revolving course, but at another epoch, when the cycles have at length reached the measure of his allotted time, he lets it go. From God, the Constructor, the world received all that is good in Him, but from a previous state came elements of evil and unrighteousness, which thence derived, first of all passed in the world, and were then transmitted to the animals. Now as long as the world was nurturing the animals within itself under the guidance of the Pilot, it produced little evil and great good; but in becoming separated from him it always got on most excellently during the time immediately after it was let go, but as time went on and it grew forgetful, the ancient condition of disorder prevailed more and more. ("The Statesman" trans. Harold North Fowler)

Ovid writes of an increase in warfare:

> The Age of Brass succeeded, as the third in order, after these; fiercer in disposition, and more prone to horrible warfare. (Ovid, *The Metamorphosis,* book 1 trans. FJ Miller)

The Age of Iron

There will be terrible times in the last days. People will be lovers of themselves, lovers of money, boastful, proud, abusive, disobedient to their parents, ungrateful, unholy, without love, unforgiving, slanderous, without self-control, brutal, not lovers of the good, treacherous, rash, conceited, lovers of pleasure rather than lovers of God having a form of godliness but denying its power. Have nothing to do with such people.
—2 Timothy 3:1–5

Ovid describes the present Iron Age:

> The last Age was of hard iron. Immediately every species of crime burst forth, in this age of degenerated tendencies; modesty, truth, and honor took flight; in their place succeeded fraud, deceit, treachery, violence, and the cursed hankering for acquisition. The sailor now spread his sails to the winds, and with these, as yet, he was but little acquainted; and the trees, which had long stood on the lofty mountains, now, as ships bounded through the unknown waves. The ground, too, hitherto common as the light of the sun and the breezes, the cautious measurer marked out with his lengthened boundary. And not only was the rich soil required to furnish corn and due sustenance, but men even descended into the entrails of the Earth; and riches were dug up, the incentives to vice, which the earth had hidden, and had removed to the Stygian shades. Then destructive iron came forth, and gold, more destructive than iron; then war came forth, that fights through the means of both, and that brandishes in his bloodstained hands the clattering arms. Men live by rapine; the guest is not safe from his entertainer, nor the father-in-law from the son-in-law; good feeling, too, between brothers is a rarity. The husband is eager for the death of the wife, she for that of her husband. Horrible stepmothers then mingle the ghastly wolfs-bane; the son prematurely makes inquiry into the years of his father. Piety lies vanquished, and the virgin Astræa is the last of the heavenly deities to abandon the earth, now drenched in slaughter. (Ovid, *The Metamorphosis,* book 1 trans. FJ Miller)

The virgin Astræa personified justice and purity like the Greek Dike, who disgusted with human greed, departed:

> Now comes the last age of the Cumaean song;
> the great order of the ages arises anew,
> Now the Virgin returns, and Saturn's reign returns;
> now a new generation is sent down from high heaven.

> Only, chaste Lucina, favor the child at his birth,
> by whom, first of all, the Iron Age will end
> and a golden race arise in all the world ...
> (Virgil, *Eclogue*)

St. Augustine, Dante, Abelard, and others believed that the Sybil of Cumae was prophesying the advent of Christ.

The ancient Persians shared St. Paul's prophetic view of the last days of the Iron Age. The Zend Avesta predicts a vigorous assault on the family and religion by the powers of darkness and a cosmic battle between god (Ohrmazd) and the devil (Ahriman). There will be signs in the sun and moon, darkness and gloom, earthquakes, famines, droughts, pestilence, and terrible battles, until a righteous prince arrives to restore all things.[104]

The Hindu fourth and present age, the Kali-yuga (black age or naught) is the worst of all. It is an age of discord in which human beings are both weak and evil. They call it the "Age of Quarrel and Hypocrisy."[105] Humanity and society reach the extreme point of disintegration. Social and spiritual life degenerate to its lowest point, and truth and love disappear from the earth. Confusion and spiritual decadence characterize this final stage.[106]

The Hopi fourth and present world is known as Tuwaquichi. Its color is yellowish white; its mineral is mixed. Like the Kali-yuga, this age is characterized by "man's ruthless materialism and imperialistic will." The gross appetites of the flesh predominate.[107]

In the eighth century BC, Hesiod describes the present age,

> For now truly is a race of iron, and men never rest from labor and sorrow by day, and from perishing by night; and the gods shall lay sore trouble upon them ... The father will not agree with his children, nor the children with their father, nor guest with his host, nor comrade with comrade; nor will brother be dear to brother as aforetime. Men will dishonor their parents as they grow quickly old, and will carp at them, chiding them with bitter words, hard-hearted they, not knowing the fear of the gods. They will not repay their aged parents the cost of their nurture, for might shall be their right: and one man will sack another's city. There

will be no favor for the man who keeps his oath or for the just or for the good; but rather men will praise the evildoer and his violent dealing. Strength will be right and reverence will cease to be; and the wicked will hurt the worthy man, speaking false words against him, and will swear an oath upon them. Envy, foul-mouthed, delighting in evil, with scowling face, will go along with wretched men one and all. (Hesiod, *The Works and Days* trans. Hugh G. Evelyn-White)

The rapid disintegration of civilization that we see all around us during the present conclusion of the Iron Age is not without hope. The angel Gabriel prophesied to Daniel.

And in the days of those kings the God of heaven will set up a kingdom that shall never be destroyed, nor shall this kingdom be left to another people. It shall crush all these kingdoms and bring them to an end, and it shall stand forever; Just as you saw that a stone was cut from the mountain not by hands, and that it crushed the iron, the bronze, the clay, the silver, and the gold. The great God has informed the king what shall be hereafter. The dream is certain, and its interpretation trustworthy. (Daniel 2:44, 45)

Chapter 5

The Mountain of God

The Mountain of His Holiness

Great is the Lord and greatly to be praised in the city of our God, in
his holy mountain. Beautiful for situation, the joy of the whole earth
is Mount Zion, on the sides of the north, the city of the great King.
—Psalm 48:1, 2 (NKJV)

It is stated many times in the Bible that the Creator's eternal dwelling is on
Mount Zion: "The Lord of hosts, who dwells on Mount Zion" (Isaiah 8:18).
"The Lord will reign over them in Mount Zion now and evermore" (Micah
4:7). No other mountain can be compared to it: "O splendid many-peaked
ranges! Well may you look with envy on Mount Zion, the mount where God
has chosen to live forever" (Psalm 68:16). In world mythology, this mountain
is described as "the seat of the gods," "the highest point on earth," and "the
place where the Creation began."[1] From antiquity, this holy mount was the
model for numerous religious buildings and earthworks.

The ridge in Jerusalem called Mount Zion could never be described in
such terms. In Isaiah 14:13, it is the coveted place of Lucifer, who said, "I will
ascend into heaven, I will exalt my throne above the stars of God: I will sit also
upon the mount of the congregation in the side of the North." The heavenly
Jerusalem is situated on Mount Zion: "But you have come to Mount Zion
and to the city of the living God, the heavenly Jerusalem and to innumerable
angels in festal gathering. For you have not come [as did the Israelites in the

wilderness] to a [material] mountain that can be touched, [a mountain] that is ablaze with fire" (Hebrews 12:18a, 22 AMP).

Every tribe has its own name for the Mountain of God. The Gikuyu people of Kenya say their holy peak was once God's resting place. The Creator made a mountain called Keré Nyaga (Mountain of Brightness or Mystery). It was a sign of His wonders and God's dwelling place on earth. The people turn toward it when praying, and they offer sacrifices there.[2] According to the Siberian Altaic tribes, the Creator, the Over-God, once sat on a golden throne in a high palace on a golden mountain in the middle of the sky. He lowered His holy mountain Altai (of iron) onto the earth.[3]

The Aymara people of Bolivia believe their 21,000-foot-high Mount Illimani, which looks down on the city of La Paz, is the dwelling place of the Creator Viracocha.[4] The Japanese Kojiki calls the Mountain of God Ame-No-Kana-Yama (Heavenly Metal Mountain). The Aino, Japan's oldest inhabitants, call their holy hill Kogane-yama. They say it is a remnant of a very tall mountain whose top extended into the skies, and upon its summit was a "a great house of fog."[5]

The Mountain of God, Stele of Naram-Sin c. 1250 BC
Encyclopedia Britanica (1910)

The Marriage of Heaven and Earth

The Lord will take delight in you, and your land will be married.
—Isaiah 62:4b

In North America, the Pawnee, Sioux, and Omaha tribes celebrate the marriage of heaven and earth as one of their sacred mysteries.[6] According to the Zuni, earth and sky once lay in union with each other until they were forcefully separated.[7] The Maori of New Zealand tell of the grief caused by the severing of the sinews that united heaven and earth.[8] The Bushongo of the Congo say, "Heaven and earth once lived united like husband and wife; but Heaven went off in displeasure."[9] The memory of the sacred marriage and great divorce at the Flood is universal.

The Chinese sages tell of the union of heaven and earth before the Flood, and the breaking off of communication between them.[10] According to Kwang-Tse:

> Heaven and earth are the father and mother of all things.
> It is by their union that the body is formed; it is by their
> separation that a (new) beginning is brought about.
> (*The Texts of Taoism* volume 2 trans. James Legge)

Yao Tiandi is described as "a personified mountain towering over his watery domain for countless years." He "heaped up the earth on a high base," but a terrible flood came during His reign.[11]

The Hindu texts speak of two worlds once united. From his comprehensive research of the Hindu texts, the late University of Chicago professor Mircea Eliade concluded that every Indian marriage is a reenactment of the original cosmic marriage. "I am heaven," declares the husband and "you are earth." Every husband and wife is a living illustration of that original sacred union.[12]

Geologists have determined that the earth is not a perfect sphere but somewhat pear shaped. Christopher Columbus described it this way:

> The earth is not perfectlye rounde; but that when it was
> created, there was a certeyne heape reysed thereon, much
> hygher than the other partes of the same. (Pietro Martire
> d'Anghiera Christophorus Colonus, the Admyrall
> *Star Names: Their Lore and Meaning*, Richard H. Allen 1889)

Columbus called the mountain "Paria," saying it contained Paradise.[13]

With heaven and earth's rotation in sync, the geographical site directly below the Throne of God was greatly distorted by the immense gravitation and electromagnetic force of the larger body. Astrophysicist Alan Friedman explained this phenomenon. If two celestial bodies join in marriage and go around together, normally their spins would stay the same, but if one is larger and heavier, its gravitational pull would slow down the other one. A bulge on the smaller body would be attracted to and face its larger mate, and soon their revolutionary and rotational periods would be identical.[14] There was only one ocean, and all inhabited land was contiguous; the earth's continental plates were held together in a Pangaea-like super-continent, what geophysicist and meteorologist Alfred Wegener suggested in 1912. In fact, the sacred mountain of Macedonia reflected this in its very name: Mount Pangaios (the whole earth).[15]

A Flat Earth

Every valley will be raised up, every mountain and hill made low;
the rough ground will become level, the rugged places a plain.
And the glory of the Lord will be revealed, and all mankind
together will see it. For the mouth of the Lord has spoken.
—Isaiah 40:4, 5

The Creator radically changed the contours of the land at the Flood. According to many tribal traditions the surface of the earth was once relatively flat with only gently sloping hills. The Persians say that the earth in its original perfect state had no mountains or valleys, and everything was in harmony and peace.[16] The Inuit of Alaska speak of an earth with perpetual sunshine, no rain, and no mountains, near or far.[17] There were no great mountain ranges, such as the Rockies, Alps, Andes, and Himalayas.

The tribes of the Pacific Northwest say there were no mountain peaks, only a great grassy plain. The Great Spirit had covered it with a huge cloud, but after the cloud went away, the land changed, and huge mountains rose on all sides.[18] Primal heroes and fisher kings, such as Maui of Polynesia and Qat of Melanesia, are remembered as "pulling or fishing up the land."[19] The Inca Creator Kon Tiki Viracocha "lowered the mountains and raised the valleys by the power of his will and word" and gave the people abundant bread and fruit. Viracocha was also the god of the waters who vanished.[20]

The natives of Collao in the central Andes tell of one with such power that he changed the hills into valleys and the valleys into hills.[21] The Mayan Popul Vuh states that Seven Macaw, whose reign ended at the Flood, was able to raise and level mountains in a single day.[22] The Persians blame Ahriman (the devil) for making the earth shake and mountains grow,[23] but Ahura Mazda will have the final victory.[24]

During the Australian dreamtime, before they disappeared, the supernatural beings changed the physical landscape. It was essentially a remolding and transformation of preexistent material. The Aranda tribes call this the Alcheringa Time.[25] Among the western Aranda and Loritja tribes, it is said that the dreamtime heroes had access to the sky world by means of a mountain, but their "Sky-hero" caused the mountain to sink, and they now had to remain on earth.[26]

Geologists have no universally accepted explanation for orogenesis (mountain building). Few believe that mountains arose in the memory of human beings, but these uplifts were a worldwide phenomena[27] (Psalm 104:4–8, Genesis 7:11). Whitcomb and Morris write:

> It is extremely interesting, in light of the biblical suggestion of uplift of the lands at the conclusion of the deluge period, to note that most of the present mountain ranges of the world are believed to have been uplifted (on the basis of fossil evidence) during the Pleistocene or late Pliocene ... supposed to represent the most recent geological epochs, except that of the present, and since nearly all the great mountain areas of the world have been found to have fossils from these times near their summits, there is no conclusion possible other than that the mountains (and therefore the continents of which they form the backbones) have all been uplifted essentially simultaneously and quite recently. (Whitcomb/Morris, *The Genesis Flood*)

The Tent of the Lord

I long to dwell in your tent forever and take
refuge in the shelter of your wings.
—Psalm 61:4

The tidal bulge explains part of the mystery of Mount Zion, but all traditions describe a mountain reaching to heaven. In the previous chapter, we spoke of the vapor canopy that supplied the prodigious amount of water required for a worldwide flood. How it was supported has been a continuing problem for canopy scientists.[28] What mechanism held it in place? Repeatedly, the Bible tells us the shape of the preflood atmosphere. The canopy was not a thick cloud cover like sizzling Venus but a thermal tent over the stationary earth. "He stretches out the heavens like a canopy, and spreads them out like a tent" (Isaiah 40:22b). "In the heavens he has pitched a tent for the sun" (Psalm 19:4). "Wrapped in light as with a garment. You stretch out the heavens like a tent" (Psalm 104:2).

The triangle has symbolized communication between heaven and earth from antiquity.[29] The Hebrew letter D *dalet* ד (Paleo-Hebrew △ Greek *delta* ⚠) was originally a triangle meaning "door." The Chinese word *hui* 会, originally △, means "union."[30] God's tent was suspended over the antediluvian world, and the earth was held firmly in His embrace. In any Hebrew dictionary, the first meaning of the second letter of the alphabet, *beit* ב, is "tent." According to the Midrash, all the children of Adam once sat under God's tent, and its collapsing waters caused the Flood.[31]

His Word and His holy arm accomplished this. "I am the first and I am the last. My own hand laid the foundations of the earth, and my right hand spread out the heavens; when I summon them, *they all stand up together*" (Isaiah 48:12b–13). Josephus wrote of this marvel,

> On the second day, He placed the heaven over the whole world, and separated it from the other parts; and He determined it should stand by itself. He also placed a crystalline firmament round it, and put it together in a manner agreeable to the earth, and fitted it for giving moisture. (Josephus, *Antiquities of the Jews*, book 1, chapter 1)

The Sumerian hymns speak of its shade and refreshment.

> The standard that reaches out from the Abzu
> that has been made into a canopy,
> whose shade stretches over the entire earth,
> refreshing its people.
> (Samuel N. Kramer and John Maier, *Myths of Enki the Crafty God*)

The *Rg Veda* recalls heaven's suspended waters:

> Thou art the counterpart of earth, the Master of lofty heaven
> with all its mighty Heroes: Thou hast filled all the region
> with thy greatness: yea, of a truth there is none other like
> thee. Whose amplitude the heaven and earth and waters of
> midair have never reached. (Book 1, Hymn 52,
> Ralph TH Griffith, trans, *Hymns of the Rg Veda* London
> 1889)

The cloud of glory on Mount Sinai was a type of the Lord's original tent. "The glory of the Lord settled on Mount Sinai, and the cloud covered it for six days; on the seventh day he called to Moses out of the cloud" (Exodus 24:16). A cloud of glory later covered the tent of the congregation. "Then the cloud covered the tent of meeting, and the glory of the Lord filled the tabernacle" (Exodus 40:34). The cloud of glory was the indication of His presence. While the original canopy was stationary (James 1:17 AMP), the latter moved with the camp:

> Whenever the cloud lifted from over the tent, then the
> Israelites would set out; and in the place where the cloud
> settled down, there the Israelites would camp. (Numbers 9:17)

A tent-shaped canopy is a beautiful symbol of Christ. "… The true tent that the Lord and not any mortal, set up" (Hebrews 8:2b NRSV). The cloud canopy and bulging lithosphere appeared as one cone-shaped body to the people. No matter where they stood on earth, the people looked up into the Kingdom of God shining forth from His lofty boreal peak. The canopy waters now make up the earth's oceans and frozen poles, but earth's strong magnetic field remains.

The Sweat Lodge of the Great Spirit

The house of the wicked is destroyed, but the tent of the upright flourishes.
—Proverbs 14:11

In their dwellings and ceremonies, Native Americans have preserved the memory of the tent of the skies. The Wintu of California tell of the great sweathouse of Olelbis (dwelling or sitting on high), who sees everything. Called Olelpanti Hlut, it was the model for all others.[32]

> It stood there in the morning dawn, a mountain of beautiful flowers and oak tree branches; all colors of the world were on it, outside and inside. The tree in the middle was far above the top of the house, and filled with acorns; a few of them had fallen on every side. The sweathouse was placed there to last forever, the largest and most beautiful building in the world above or below. Nothing like it will ever be built again. (Jeremiah Curtin, *Creation Myths of Primitive America*)

The sweat lodge ceremony is a sacred ritual for healing, purification, prayer, and endurance. It is believed that sickness or ill luck will strike anyone who builds the house irreverently. Its entrance should face a lake, stream, or sacred fire. Among the Sioux, the sweat lodge is where preparations for the sun-dance ceremony are held, an ancient celebration of the Creation. Pails of water and steaming hot stones are brought into the lodge, and water is poured over them. The bather prays silently and comes through with renewed vigor.[33] When the religious significance of the sweat bath faded, as in the Swedish sauna and Turkish bath, the practice still continued for health reasons. Elsewhere, purificatory bathing in sacred rivers replaced the sweat bath.

The Olympian Peaks of Antiquity

For the mountains shall depart, and the hills be removed, but
My kindness shall not depart from you nor shall My covenant of
peace be removed," says the Lord, who has mercy on you.
—Isaiah 54:10 (NKJV)

Though scattered and divided by language, the people never forgot the Creator's holy mountain. Every tribe embraced its own local hill as the sole resting place of their God untouched by the waters of the Flood.[34] In scaling its heights, they felt close to Him and their ancestors whose road led up that mountain. The Assyrian expression for dying was "grappling oneself to the mountain."[35] Pilgrimages were made to these "high places" barefoot or on the knees, and although the ancient rites go on in many places, the true meaning has been lost.

In flatter areas, the people sought to duplicate the holy mountain in their temples, palaces, tombs, and earthworks, all constructed with great effort. Mounds of stones, called cairns, were set up in honor of the ancestors, and pilgrims added a stone as they passed. Earth mounds or barrows were used as shrines and tombs. The Kawai and Inu shrines in Japan, Silbury Hill in England, and Cahokia in Illinois are examples of man-made mounds.

Greece: Mount Olympus

Then I looked, and there before me was the Lamb, standing
on Mount Zion, and with him 144,000 who had his name
and his Father's name written on their foreheads.
—Revelation 14:1

Olympus is a pre-Hellenic word meaning "mountain." As Homer describes it:

Olympus, where they say there is an abode of the gods, ever unchanging: it is neither shaken by winds nor ever wet with rain, nor does snow come near it, but clear weather spreads cloudless about it, and a white radiance stretches above it. (Homer, *Odyssey,* trans AT Murray 6)

From its peak, Kronos ruled over all.

I shall come again to give you sweeter glory yet,
Finding in my swift chariot a way of words
As I approach the sun-filled hill of Cronos.
(Pindar, Olympian Ode 1)

Mount Olympus was the first Acropolis. When it collapsed, the sky fell. Hesiod describes the scene, which made all tribes live in fear of it falling on them again:

> It absolutely would have seemed as if the earth and the wide heaven above her had collided, for such was the crash arising as earth wrecked and the sky came piling down on top of her. (Theogony trans. Hugh G. Evelyn-White)

Mount Ida was Troy's holy hill in the north, upon whose summit Zeus dwelt. There he held council to decide the fate of humanity. His throne, covered in blue, had seven steps, each enameled with a different color of the rainbow.[36] He was said to be present on every hill and was worshipped on high places.[37]

Mesopotamia: Mount Mashu

> Yet have I set my king upon my holy hill of Zion.
> —Psalm 2:6

An excerpt from the Sumerian creation epic tells how An ✳ (god, star) separated heaven and earth:

> After heaven had been moved away from earth,
> After earth had been separated from heaven,
> After the name of man had been fixed;
> After An had carried off heaven,
> After Enlil had carried off earth …
> (Samuel Noah Kramer, *Sumerian Mythology*)

In primordial times, An lived on the Du-ku (holy hill), the cosmic mount where the fate of humanity was determined.[38] Enlil, the supreme god of Nippur (navel), was called "the Great Mountain" and was worshipped in his E-kur (Mountain house).[39]

The holy hill of Akkad was Kharsak Kur, dazzling with gold and precious stones:

> O mighty mountain of Bel, Im-Kharsak, whose head rivals heaven, whose root is in the holy deep! Among the

mountains like a strong wild bull it lieth down. Its horn like the brilliance of the sun is bright. Like the star of heaven, it is filled with sheen. (Akkadian hymn[40])

Babylon called itself "the home of the luminous hill," Mount Mashu.[41] Its cuneiform symbol is ✝. The Ziggurat was a miniature Mount Mashu, an antechamber where the gods conversed with humans. It was called "a likeness of what was made in Heaven."[42] With seven ascending stages, its sides faced the four directions. A tablet in the Louvre records the first level as black and dedicated to Saturn. The walls leaned in, all lines leading the eye inward and upward.[43] The north point with four gates contained the *baldachin,* a golden canopy emblazoned with stars. The king was ritually married to his land at the new year in the topmost shrine symbolizing the marriage of Heaven and earth.[44]

India: Mount Meru

Exalt the Lord and worship at his holy hill; for the Lord our God is Holy.
—Psalm 99:9

In Hindu cosmology, the king of mountains is Mount Meru, over which shines the polestar. Home of the gods and Gandharvas, it is unapproachable by sinful men:

> It stands kissing the heavens by its height and is the first of mountains. Ordinary people cannot even think of ascending it. It is graced with trees and streams, and resounds with the charming melody of winged choirs. Once the celestials sat on its begemmed peak in conclave. (*The Mahabharata* book 1 trans. Kisari Mohan Ganguli)

Among its titles are Hemadri (golden mountain) and Karnikachala (central mountain).[45] It had the shape of a pure seven-sided pyramid, and each face was a different color of the rainbow.[46] It was estimated to be 844,000 miles high and 320,000 miles wide. The celestial river Ganga encircled its summit subdividing into four rivers.[47] The mountain influenced societal divisions. The highest caste—the priests—lived in the middle of the town, having the highest houses. The other castes—soldiers, tradesmen, and laborers—lived in their various lower zones.[48]

The Buddhist World Mountain is Sumeru, attained by adding the prefix *su* (beautiful) to Meru. Related are the Mongolian Sümmer Sola, the Buryat Sumu, Uralo-Altaic Semeru, Chinese Siumi, and Japanese Shumi. The Buddhist temple is an image of Sumeru, and the canopy is one of the Eight Treasures.[49] Mount Kailash, a place of pilgrimage sacred to the Bons, Hindus, Jains, and Buddhists, is linked with Sumeru. They remember it as the world navel and palace of the gods.[50]

China: Kunlun

But he who takes refuge in me shall possess the
land and shall inherit my holy mountain.
—Isaiah 57:13b

The Taoist sacred mount is Kunlun or Kuenlun (Pearl Mountain) rising to the polestar. It was the dwelling place of the immortals and blessed dead. It had four rivers, twelve jeweled towers, and a jade tree of life. The mountain symbolized the greatness of the emperor, the son of heaven.[51] Representative buildings on earth had four cardinal staircases approaching the Temple of Heaven at the capital. Royal tombs were built on mountain peaks with the head of the ruler, facing north. Princes were buried on the hills while the lower classes were buried on the plains.[52] The most famous of these peaks is Mount T'ai in Shantung. A stone from there is said to ward off demons.[53]

Mexico: Colhuacan

It is as if the dew of Hermon were falling on Mount Zion. For
there the Lord bestows his blessing, even life forevermore.
—Psalm 133:3

The Aztecs said it was Quetzalcoatl, the first being of creation, who "lifted up the waters from the earth to the sky."[54] Colhuacan in the far north was their place of origin. The sky to them was the ceiling of a towering edifice, sea and sky being of one substance. Religion was consumed with avoiding the collapse of the sky waters a second time, annihilating humanity. High mountains were holy, especially those with caves. The people made arduous pilgrimages there for worship and sacrifice. Popocatepetl, with its cave and shrine, was one of the holiest.[55] The Hill of the Star (Cerro de la Estrella) near Mexico City is the ancient site of the New Year and New Fire ceremony.[56] Step pyramids with the god's house on top (Teocalli) were built as centers of ritual in flatter areas, all oriented to the four directions.

Persia: Mount Alburz

Who may ascend the hill of the Lord? Who may stand in his holy
place? He who has clean hands and a pure heart, who does not
lift up his soul to an idol or swear by what is false. He will receive
blessing from the Lord and vindication from God his Savior.
—Psalm 24:3, 4

Mount Alburz (white) is the Persian holy mount. It was said to have taken eight hundred years to grow to "the utmost limit of the sky," and at its summit the gods resided. Mount Taera is the central peak of Alburz. It is called "the chief of summits."[57] Tall peaks in Persia were seen as its representations. The Zend Avesta calls it Hara Berezaiti (High Watchpost),

The bright mountain around which the many stars revolve,
where come neither night nor darkness no cold wind and no
hot wind, no deathful sickness, no uncleanness made by the
Daêvas. (Rashnu Yasht, Zend-Avesta)[58]

Its roots spread under the earth, holding it together. The garden paradise of King Yima was situated on its summit.[59]

Egypt: The Primeval Hill

O Lord, who may abide in your tent? Who may dwell on
your holy hill? Those who walk blamelessly, and do what
is right, and speak the truth from their heart.
—Psalm 15:1, 2

In Egyptian cosmology, earth joined the sky at its northernmost point, the summit of the Primeval Hill.[60] All the priests of the great cult centers claimed their temple stood on it.[61] A seven-step staircase △ represented it on the Pyramid Texts. These were considered *asket pet* (steps to heaven).[62] Asar (Osirus) is seen seated on a throne of seven steps and called "the Pyramid" in the texts.[63] Incorporating the cross, triangle, and square ⊠, the pyramid represents the Primeval Hill, with its Foundation Stone on top.

The Great Pyramid of Giza is the only survivor of the seven wonders of the ancient world. Its architects exhibited unique scientific knowledge and engineering skill. Its square base is aligned to the cardinal points based on true north, an accuracy never equaled in any other edifice.[64] It is at the apex of the Nile Delta, and its meridian divides it into equal halves. Its southern border is exactly seven terrestrial degrees in length. Pi π is built into it.[65] Its site intersects the earth's longest land meridian and its longest land parallel. Unlike inferior copies, it contains no pharaoh's name or mummy, and it is the only one to have chambers above ground.[66]

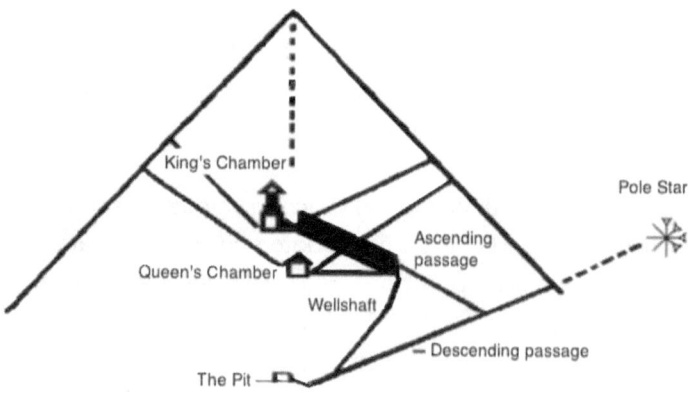

The Pyramid of Giza, 2560 BC (ES)

Physicist and author John Zajac has pointed out that the walls of the Queen's Chamber are encrusted with salt half an inch thick, which he concludes is a remnant of the Flood. Its entrance is in the middle of its north face, and its long passageway out points to the polestar. All four faces, each 440 cubits at its base, were originally covered with 144,000 white casing stones so brilliantly polished they could be seen in Israel hundreds of miles away. Vandals stole them in the last thousand years, and its famous capstone is missing.[67]

The pyramid casts no shadow at noon on the vernal equinox. If one views sunrise from there that day, it rises directly over Bethlehem. The Pyramid of Giza (border) fits Isaiah 19:19 exactly. "On that day there will be an altar to the Lord in the *center* of the land of Egypt, and a pillar to the Lord at its *border*. It will be a sign and a witness to the Lord of hosts in the land of Egypt; when they cry to the Lord because of oppressors, he will send them a savior, and will defend and deliver them. The Lord will make himself known to the Egyptians; and the Egyptians will know the Lord on that day ..."[68]

Old King Cole

They will not hurt or destroy on all my holy mountain; for the earth will be full of the knowledge of the Lord as the waters cover the sea.
—Isaiah 11:9

An early Latin name for the God of Heaven was Coelus, regent of Coelum, the former sky that vanished. Column, ceiling, and conceal derive from this root. The Latin *colles* means the same thing as *mons*. The northern hills of Rome are the Collis Quirinalis, Collis Viminalis, and Collis Esquilinis.[69] The Latin root *col* is a cognate with the Hebrew *kol* (whole or all), found in America in Colhua, a name for all of Mexico, and Colhuacan, the mountain of the gods.[70] For the Toltecs, Zapotecs, Mixtecs, Aztecs, and others, Colhuacan is a "mountain far to the North, a place of origin."[71]

There are Cole's hills all over the British Isles. *Coel* in Welsh is an "omen" or "belief." Coelcerth is a "beacon or bonfire."[72] Old King Cole was the beneficent king of the golden age, an alias for the "Lord of All." He blew smoke to the four winds:

King Cole was king before the troubles came,
The land was happy while he held the helm ...
Men hear him on the downs, in lonely inns,
In valley woods or up the Chiltern Wold.
(John Masefield, "King Cole")[73]

The Eternal Mount Zion

Those who trust in the Lord are like Mount Zion,
which cannot be removed, but abides forever.
— Psalm 125:1 (NKJV)

When Jesus wanted to commune with His Father, He withdrew to a mountain (Luke 6:12, 9:28, John 6:3, 15). In Genesis 17:1, God introduces Himself to Abraham, and later to Jacob, as El Shaddai. "I appeared to Abraham, to Isaac, and to Jacob as God Almighty [El-Shaddai], but by my name the Lord [Yahweh—the redemptive name of God] I did not make Myself known to them [in acts and great miracles]" (Exodus 6:3 AMP). El Shaddai is not found anywhere outside the five books of Moses except in Job 6:4. No one has been certain what the name means.

In most translations, this name is rendered "God Almighty," but it implies more. The title is associated with the Akkadian word *shadû* (mountain) as Abraham came from Mesopotamia.[74] This same root is present in the English words *shade* and *shadow* (Old English *sceadu*). The Hebrew word *shad* means "a woman's breast." Mount Zion is "the Bosom of the Father" (Isaiah 66:10–13, John 1:18). The related Egyptian word *shet* means "to nurse." It is comforting to use the Lord's original title. "You who live in the shelter of the Most High, who abide in the shadow of El-Shaddai" (Psalm 91:1).

Israel, like other nations, looked to its own hills. "The people were sacrificing at the high places, because no house had yet been built for the Name of the Lord. Solomon loved the Lord, walking in the statutes of his father David; only he sacrificed and offered incense at the high places" (1 Kings 3:2–3). Later, idols were set up (Isaiah 57:7, 8 AMP) In the eighth century BC, King Hezekiah "removed the high places" (2 Kings 18:4). King Josiah did the same. "Josiah brought all the priests from the towns of Judah and desecrated the high places … where the priests had burned incense" (2 Kings 23:8). Jeremiah tirelessly preached against them. "Surely the idolatrous commotion on the hills and mountains is a deception; surely in the Lord our God is the salvation of Israel" (Jeremiah 3:23).

In the last days, Mount Zion will be restored: "I will go before you and will level the mountains. I will break down gates of bronze and cut through bars of iron" (Isaiah 45:2; see Revelation 16:20). "It shall come to pass in the last days, that the mountain of the Lord's house shall be established in the top of the mountains, and shall be exalted above the hills; and all nations shall flow into it. And many people shall go and say, 'Come, and let us go up to the mountain of the Lord'" (Isaiah 2:2, 3a).

Chapter 6

The Powerful Names of God

The Search for the Holy Tongue

Now the whole world had one language, and a common speech.
—Genesis 11:1

Have you ever wondered why we say, "Ow!" when we get hurt, and cry "Awa! Awa!" when we come into the world? *Wawa* means "baby" in Chinese. Why do we say, "Wow!" "Yow!" or "Yahoo!" when we get excited? These exclamations are all variations of the holy, life-giving name of YHWH. We call out to Him in our states of distress and exhilaration, even when we don't know it. People of the Snohomish tribe of Puget Sound say "Ya-hoh" when lifting something together.[1] Sailors cry "Yo-ho" when hoisting the sail, and thousands of years ago, the Egyptians referred to the God of Moses as Yaho Sabaho (YHWH of Hosts) and Ab-Iaho (Father YHWH).[2]

The first eleven chapters of the Bible record our common history. We all migrated from Babel in Iraq about six thousand years ago. Until the confusion of tongues, we all worshipped one God and spoke to Him and each other in the same language spoken by Adam, Eve, and Noah. We know this from Genesis and the testimony of widely diverse cultures. Speech was a gift given to us by our Creator in the beginning.

According to the Sumerian residents of the city of Nippur, Enlil (Lord of the Wind) alone was prince of the sky:

He is the organizing element that marks the directions in space; but at the same time he is breath, he is word, he is creative speech … and his word is holy. (Yves Bonnefoy, *Mythologies*)

All the people praised Him in one common tongue
The whole universe, the people in unison, to Enlil in one tongue gave praise. (Samuel Noah Kramer, *Sumerian Mythology*)

The Roman author and scholar Hyginus (c. 64 BC – AD 17) wrote,

For many ages men lived under the government of Jove without cities and without laws, and all speaking one language. (*Fabulæ* 143)

The Quichés of Guatemala say that the first generations of humans spoke one common tongue. All races lived together in peace, and all prayed to the heart of Heaven.[3] The indigenous peoples of the Arctic regions call the Great Spirit Raven, who created man from clay and made a wife for him. They became parents of the human race. At first, everyone spoke the same language, and their words were powerful. Whatever the people said, happened.[4] The Yoruba of Nigeria say that humans all spoke one tongue at first, but the supreme god, Oluron, gave them different languages after they became ungrateful.[5]

What happened to this common tongue? The answer is in Genesis 10 and 11. The descendants of Noah became the nations. "These are the families of the sons of Noah, after their generations, in their nations: and by these were the nations divided in the earth after the flood" (Genesis 10:32). The Flood left the people afraid and disoriented. Although they had the rainbow pledge, every time it rained, they feared a recurrence. When the Lord commanded the people to repopulate the earth, they refused to obey Him, wanting to remain secure where they were. The firmament had collapsed, and they believed they could replace it with high towers.[6]

Josephus describes how Noah's three sons tried to convince the people to leave the mountain:

Now the sons of Noah were three—Shem, Japheth and Ham, born one hundred years before the deluge. These first of all descended from the mountains in the plains and fixed their habitations there; and persuaded others who were greatly afraid of the lower grounds on account of the floods and so were very loath to come down from the higher place, to venture to follow their examples. (*Antiquities,* book 1, chapter 4)

As often happens, the people followed the wrong man. Rejecting the wise rule of Shem, an ancient ancestor of Jesus, they chose to follow Nimrod, grandson of Ham (Genesis 10:8–10). He was called "a mighty hunter in the sight of the Lord" (Geneses 10:9). The name Nimrod derives from the Hebrew word *marad* (rebel). Under his leadership, the people began to resent God. They complained about the destruction of their ancestors and supposed the prosperous life they enjoyed was due to their own efforts.[7]

Building the Tower of Babel on Mesopotamian cylinder seal.
Artwork courtesy of Ava Raha.

Josephus wrote of Nimrod:

He gradually changed the government into tyranny—seeing no other way of turning men from the fear of God, but to bring them into a constant dependence upon his power. He also said he would be revenged on God, if he should have a mind to drown the world again; for that he would build a tower too high for the waters to be able to reach, and that he would avenge himself on God for destroying their forefathers! Now the multitude was very ready to follow the determination of Nimrod, and to esteem it a piece of cowardice to submit to God: and they built a tower, neither sparing any pains nor being in any degree negligent about the work. (*Antiquities,* book 1, chapter 4)

Nimrod and his followers modeled the project on the Mountain of God. On the plain of Shinar (Sumer), "They said, 'Come, let us build ourselves a city with a tower that reaches to the heavens so that we may make a name for ourselves; and not be scattered over the face of the whole earth'" (Genesis 11:3, 4). They called their city Babel and built the tower as a terraced pyramid. Its completion became an obsession. If a worker fell off and died, they paid little attention, but if a brick fell and broke, they cried.[8] According to the Midrash, an idol with a sword was placed on top in a posture to wage war on God.[9]

The Confusion at Babel

The Lord came down to see the city and the tower that men were building. The Lord said, "they all have one language; and this is only the beginning of what they will do; nothing that they propose to do will now be impossible for them. Come, let us go down and confuse their language there, so that they will not understand one another's speech. So the Lord scattered them from there over the face of all the earth, and they stopped building the city." It was called Babel- because there the Lord confused the language of the whole world. (Genesis 11:5–9a)

The Bible does not say everyone spoke in a new tongue but that the Lord confused the language. Babel (gate of god) came to mean, "to speak incoherently" or "babble." In the Mayan texts, "A tower which was planned to reach the heavens was destroyed because of a confusion of tongues among its architects."[10] Babal in the Mayan tongue means "confusion." It is strikingly similar to the Hebrew word *balal* (to mix, to confound). The Sumerian texts agree.

The Lord of Eridu, endowed with wisdom,
Changed the speech in their mouths, put contention into it,
Into the speech of man that (until then) had been one.[11]

According to the Karens of Burma,

> In the days of Pan-dan-man, the people determined to build a pagoda that should reach up to heaven … When the pagoda was half way up to heaven, God came down and confounded the language of the people, so that they could not understand each other. Then the people scattered. (JG Frazer, *Folklore in the Old Testament* 150)

In an account from the Swinomish of Washington State, three sisters built a very high house, because they wanted to go up into the sky to see the Changer. By the time it was finished, they could not understand each other. That's why so many different languages exist now. They say the Changer rearranged the landscape after the Flood. A local mountain is named after the three sisters.[12]

> The Lohi of Admiralty Island tell of an ancient chief who said, Let us build a house as high as heaven. So they built it, and when it nearly reached the sky, there came to them from Kali a man named Po Awi, who forbade them to go on with the building. Said he to Nuikiu, 'Who told you to build so high a house?' Nuikiu answered, 'I am master of our people the Lohi. I said, 'Let us build a house as high as heaven. If I had my way, our houses should have been as high as heaven. But now, thy will is done, our houses will be low.' So saying he took water and sprinkled it on the bodies of his people. Then was their language confounded; they understood not each other and dispersed into different lands.[13]

The Siberian Ostiaks associate different tongues with the north wind:

> A strong North wind blew without ceasing for seven days scattering the people far from one another. For this reason they began, after the flood, to speak different languages and to form different peoples. (Uno Holmberg, *Finno-Ugric Mythology*)

From India, we have this account:

In the days of old, the descendants of Ram were mighty men, and growing dissatisfied with the mastery of the earth, they aspired to conquer heaven. So, they began to build a tower, which should reach up to the skies. Higher and higher rose the building, till at last the gods and demons feared lest these giants should become the masters of heaven, as they already were of earth. So they confounded their speech, and scattered them to the four corners of the world.[15]

The Tower of Babel by Pieter Bruegel the Elder

How the Tower Was Destroyed

But at your rebuke the waters fled, at the sound
of your thunder they took to flight.
— Psalm 104:7

The Bible does not say how God destroyed the tower, but according to the Talmud, the uppermost third was destroyed by fire, one third by earthquake, with one third remaining.[16] Many accounts describe an immense thunderbolt. According to Josephus, the Lord also destroyed Sodom and Gomorrah this way.[17] Ovid recounts a Roman version:

Jupiter assumed the thunderbolts after the giants dared attempt to win the sky; at first he was unarmed. (Ovid, *Fasti,* book 3)

In Iraq, building high structures is still opposed as being "of Satan."[18] There are similar American accounts. According to the Mexican Codex Vaticans, Xelhua went to Cholula to construct a giant pyramid as a refuge from another flood. When it had reached a towering height, lightning from heaven destroyed it.[19] The Apaches also recall destructive thunderbolts but confuse Noah and Nimrod:

> The first days of the world were happy and peaceful; then came a great flood from which Montezuma and the coyote alone escaped. Montezuma then became very wicked, and attempted to build a house that would reach to heaven, but the Great Spirit destroyed it with thunderbolts.
> (HH Bancroft, *Native Races of the Pacific States*)

In Southeast Asia, lightning has been regarded as a punishment for putting forbidden things together.[20] A great storm, which destroyed a tower, is part of their tradition:

> All the people lived in one large village and spoke one tongue. They set about building a tower to reach the moon. At last when the tower was almost completed, the Spirit in the moon enraged, raised a fearful storm, which wrecked it. (Sir James George Scott, "Indo-Chinese Mythology" Volume 13, *Mythology of All Races*)[21]

According to rabbinical sources, the people of Babel were not destroyed like the flood generation. The Lord showed them mercy, for they had lived together in peace and never robbed one another.[22] He only divided and scattered them. As they migrated to different parts of the world and established colonies, the common heritage of humanity was fragmented. Each tribe now had a different name for the Creator, the Garden of Eden, Adam, Eve, Cain, Abel, Enoch, Noah, Shem, Ham, Japheth, Nimrod, and so on.

The Foundation of All Speech

The Name of the Lord is a strong tower. The
righteous runs into it and is safe.
— Proverbs 18:10

The early rabbis called our parent language "the holy tongue," for with it, the
people could accomplish anything. "They have all one language; and this is
only the beginning of what they will do; nothing that they propose to do will
now be impossible for them" (Genesis 11:6). Why was it so powerful? The
Zohar offers this insight:

> Its chief characteristic was, it enabled everyone to express
> himself clearly and unmistakably in terms exactly
> corresponding to his thoughts, wishes and intentions,
> otherwise they were not understood and comprehended
> by the heavenly powers. Thus it came to pass that, by
> confusion of their speech, their power resulting from union
> of will and purpose was destroyed and nullified. Note that
> words of the holy language are understood by celestial
> beings, who when hearing them are impelled to assist and
> help those who utter them, otherwise they pay no heed or
> regard to them. This now occurred to the builders of Babel,
> who on ceasing to speak the holy tongue, lost power, and
> ability to carry out and execute their design and therefore
> left off building the city. (*Zohar, the Book of Light* 1:75b)[23]

In the Old English "Rune Poem," Os (god) is the origin of all speech. In
Norwegian, Os means "river mouth."[24] As with everything else, our words
began with God. He gave our first parents the ability to praise Him in word
and song, because all needs and desires were met in the garden. Our parent
language was holy because it was permeated with the names of God. These
make up the core of all speech and are present in every language and dialect on
the face of the earth. The Lord dictated only consonants to Moses at Mount
Sinai. There were no vowels in paleo-Hebrew script.

𐤉	𐤆	⊕	𐤇	𐤉	𐤅	𐤄	△	𐤂	𐤁	𐤀
kaf	yod	tet	het	zayin	waw	he	dalet	gimel	beyt	'alef
k	y	t	h	z	w	h	d	g	b	'

𐤕	𐤔	𐤓	𐤒	𐤑	𐤐	○	𐤎	𐤍	𐤌	𐤋
taw	śin	reś	qop	ṣade	pe	'ayin	samek	nun	mem	lamed
t	ś	r	q	ṣ	p	'	s	n	m	l

Take any vowel, add it to any consonant, and you will find a name of God somewhere in the world. As the psalmist declares, "Your name, O God, like your praise, reaches to the ends of the earth" (Psalm 48:10). As God's names are eternal, the holy tongue is not dead, no matter how widely scattered or isolated. These names minister life wherever they're spoken. We find them in basic words associated with light, life, love, action, and origins. We will begin with the holy names of God in the Bible.

El, the Eternal Light

God is light and in Him is no darkness at all.
—1 John 1:5

All the light in the universe radiates from El. Even the stars cannot shine without Him (Job 9:7, 8). El begins the Sumerian *Ilu* (god), *ellu* (shining, pure), and *allilu* (powerful). El is the root of the Inca *Illa* and Greek *elaphros*. Both mean, "light." El begins *ill*uminate, the Hungarian *elet* (life), and Turkish *olmak* (to be), for "in him we live and move and have our being" (Acts 17:28). *Soul* is *ilma* in Finland, *elimo* in the Congo, and *ya'al* in the Caroline Islands. The Sumerian word for wind or breath is *lil,* which is *lelek* in Hungarian.

El begins *element*, Latin for first principle. It also begins *elect, elite,* and *eligible,* words meaning "chosen." El is the center of Hula, the sacred dance of Hawaii, the *Hulu* of New Zealand, and Hebrew *chul* (to turn in a circle). El begins the Greek *eleeo* (compassion) and Gaelic *oll* (vast). El ends the Hebrew *gadol* (great) and Gaelic *eol* (knowledge). El is the ring in *bell* and the sweetness in the French *miel* (honey). El begins Elijah (Yah is God), Elishah (God is salvation), and Elizabeth (God is my oath). El ends Ezekiel (strength of God), Joel (the Lord is God), Gabriel (God is my strength), and Immanuel (God with us), the title of Christ in Isaiah's prophecy (Isaiah 7:14).

El is the Most High worshipped on hills (Hebrew *tel*), so we find His name in the Latin *altus* (high) and Babylonian *ilatu* (zenith). El ends *excel*, the French *ciel* (heaven, sky), and German *Himmel* (heaven). El is the center of the Sanskrit *acala* (immovable mountain). As the *ul*timate and *ul*terior, we find El in many names for heaven, the Latin Elysium, the Trojan Ilium, the Greek Olympia, the Hindu Ilâvrita in the north, the Toltec Tulan, and holy sites, Ellora, Elis, Eleusis (advent), and Elephantine.

God is Elil in Sumerian, Old Irish Aillil, Hittite Ullu, Swedish Ull, Ostiak Ilem, Finnish Yumala, Altaic Tatar Bai Ulgen, and Lapp Ibmel. The Tlingits of Alaska call Him Yehl.[25] The Tehuelche of Argentina know Elal as their great hero of creation.[26] The Kooris of Australia worship Bunjil. Woolool Woolool is the Great Spirit of Wellington Rocks, New South Wales. Wullunggnari is sacred to the Kimberley people, where three stones represent the great Flood. Walguna is their tree of life.[27]

The Egyptians did not have L. They used R instead. The verb *ar* or *ari* means, "to make, create, form," and Ari is "the creative god."[28] Israel is Israar. Shalom is *sharm* (to greet, praise, salute). The Egyptian Mountain of God is Herara (Mountain of Ra) similar to the Hebrew Hariel (Mountain of El). Al in Hebrew and Ar in Egyptian both mean "to go up." The names of high priests Eli and Aaron both mean "high." Ar begins words meaning "highest," like *ar*chbishop and *ar*changel. *Ariel* (Hearth of God, Lion of God) is a title of both Christ, the Lion of Judah, and Jerusalem (Isaiah 29:1a, 7). The names Uriah and Uriel both mean "light of God" and "flame of God."

Ar is the "light" in st*ar* and the Egyptian *aur* (brilliance).[29] Ar begins words meaning ancient: *ar*chaic, *ar*chaeology, and legendary first ancestors, such as the Arcadians, Arioi (Medes), Argonauts, and Aryans. Ar also begins holy places like Arallu, Ariadan (Eridanus), Arcadia, Arionrhod, Eridu, Erin, Ur, Uruk, and so on. The oldest part of Rome, where the most ancient sacrifices were made, was the Argea.[30] Ar begins arch, ark, arthro (join), arrange, and architect.

The Great I AM

God said to Moses, I AM WHO I AM.
—Exodus 3:14a

God revealed to Moses His ineffable name, YHWH. Found 6,823 times in the Bible, it was only uttered once a year by the high priest on the Day of Atonement. It is so holy it can't be fully comprehended (Judges 13:18) but is close to the One that brings into being. It is a compound of *Yah* (I AM) and *Hwh* (life). *Yah* is the voice of ascent, source of *ja*, yea, yes, and the Chinese yú. When we say "Hi-ya!" we confess "life" (Hebrew *chaya)* over each other without knowing it. *Hayyei Olam* is "Eternal life." *Aya Aya Aya Aya* is a widespread Native American chant. Isaiah, Jeremiah, Josiah, Nehemiah, Hezekiah, and Zechariah bear His name.

YHWH יהוה reveals Himself in the letters of His name. Hebrew letters are also words. *Yod* (hand) is the first letter. The third is *vav* ו (nail). The second and fourth letters are *heh* ה (breath), which became the letter E, originally 𐤄. It later became ⊤ in South Arabic script, and ⊣ in proto-Phoenician. In Greek and Latin, it was turned to the right.[31] Its original form is a picture of the Word of God, Yeshua, who stands as the great Hero Mediator between heaven and earth. There are two Hehs in the holy name because of His two natures (lamb and lion) and two comings.

So many great words contain the name Yah, the Hebrew *yadah* (give thanks), *yashar* (upright), *yadid* (beloved), *yafa* (shine), and *yad yamin* (right hand). When we say "Hallelujah!" (Aramaic "Alleluia!"), we magnify the Lord's name four times El/El/El/Yah. The Greek form of *ya* is *eu* (good). *Eu* begins *eulogeo* (praise), *euche* (prayer), *euthutes,* (righteousness), *eusebos* (godly), and Eucharist (express gratitude). Words like *eudia* (fair weather), *eunoia* (good will), *euporia* (wealth), *eunomia* (good order), and *euphoria* (well bearing) express His nature.

The Latin equivalent of *yah* is *iu*. *Iu* is the root of *ius* (right, law), the root of justice, just, judge (*iudex*), jury, jurisdiction, injury (what is unlawful) and *iuro* (to swear an oath). *Iu* is the root of *iugam* (crossbar), *iugo* (bind together), and *iuncto* (join). *Ia* begins Ianus, the Etrusco-Roman god One, for whom January is named. Ius is cognate with the Vedic *Yos* and Avestan Yaos (integrity and mystical perfection).[32] Ya begins the Sanskrit *yah* (who), *yat* (what), and *yada* (when). The Latin *iuvenis* (youth), English rejuvenate, and Sanskrit *yuvan* (young) all originate in the One who is the fountain of youth.

In Anglo-Saxon England, Yah and El combine to form yule (wheel). When nature was cold and dead, yule celebrated the God whose kingdom resembled a wheel. This feast was a renewal of life, symbolized by the evergreen, "I am like a green fir tree. From me is thy fruit found. Who is wise, and he shall understand these things?" prudent, and he shall know them? for the ways of the LORD are right, and the just shall walk in them: but the transgressors shall fall therein. (Hosea 14:8b, 9 KJV). The yule log, believed to bless the home, is a memory of the tree of life. Yule is *jol* in Old Norse, from which the word *jolly* is derived.

The Sound of Life

The tongue has the power of life and death, and
those who love it will eat its fruit.
—Proverbs 18:21

HWH, the second name in YHWH, is the Creator's Life-breath. God made Eve (Hebrew Hava, Aramaic Hawa), "the mother of all the living" (Genesis 3:20) HWH, and its variants (VA, WA, UA, or the reverse AV, AO, OW, etc.) are in thousands of words meaning "life" and "breath." Wa in Swahili is "to be." Wuinic in Mayan is "being." Ruah is Hebrew for "spirit." In Maori, wairua is "spirit" and Manuwa ora is the "breath of life." In Java, *nawa* means "breath, life, and soul." Love begins and ends with His name (El/HV). Love is *lyubov* in Russian and *ahava* in Hebrew. HVH is the root of the Latin *avus* (ancestor), the Hebrew *ava* (father), and *avodah* (prayer). He is the *aw* in awesome, the *ow* in power, and the Mandarin *kowtow* (show reverence for).

A Powwow is the Native American sacred assembly. Wallum Olam is the creation text of the Delaware Indians. Awahoksu is the Pawnee abode of spiritual power.[33] The Yamana of Tierra del Fuego call Watawinewa "Powerful One, My Father."[34] Túwanasavi is the Hopi "Center of the Universe," the spiritual home to which they take the Hawiovi (One-Way Trail).[35] Hau-k-in is "teacher" in Mayan, and Ha hawkan are the holy men of the Oglala Sioux. Huacas are Inca sacred places and the Callawaya are holy men.

Awa is the Japanese Land of Eternity, and *uwa* means "up." Auwa is a place of divine power for the Aranda of central Australia. Kawa-auwa is the name of their tree of life. Their God, Numbakula (always existing), planted it in the middle of His sacred ground. *Awa! Awa!* is a sacred chant in northeast Australia.[36] The Lord's name begins the Polynesian *vatea* (sky) and the Icelandic Hávamál (Words of the High One).[37]

An, the God One

Hear, O Israel: The Lord our God, the Lord *is* One!
—Deuteronomy 6:4

An ✳ is a name for the Most High God in Mesopotamia symbolized by an eight pointed star or one stroke 𐎛 , as He is the god One.[38] That the patriarchs once knew God as An is seen in the names Beth-An (House of An), Bethany, Cana and Canaan. Daniel (God is my judge), Jonah (dove), Nathan (gift), Zephaniah (hidden by God), Jonathan (God has given), Hannah (grace, favor) bear this name. An appears twice in *ananim* (clouds) (Daniel 7:13). Mana is spiritual power in Polynesia. Jesus gives "hidden *manna*" (Revelation 2:17). He is merciful to the *oni* (poor) and *anavim* (meek). *Ani* means, "I am" in Hebrew. *An* in Egyptian is both God and His city (Biblical On, Greek Heliopolis). *An* also means, "beautiful, delightful."[39]

The name of God is heard all over the Americas, and the sign of God is seen in thousands of symbols. *An* is the Mayan "to be." Hunab Ku (Only god) is a name for the Creator. *An* is present in *chan* (heaven, sky), *chun* (foundation), and *tun* (360–day year). *Kuna* is His holy house or temple. According to linguist Dr. Morris Swadesh of the National University of Mexico, there is a 20 percent relationship between Hebrew and Mayan.[40] The Mayan word *matan* (given) is almost identical to the Hebrew *matan* ("giving"). The tribes of the Yucatan descended from Joktan, son of Eber (Genesis 10:25).

The Divine Name in China and Japan

Salvation is found in no one else, for there is no other *name*
under heaven given to mankind by which we must be saved.
—Acts 4:12

Early Chinese characters are pictographic, and the sign of God appears in many. *Yu*由 means germination, beginning, principle, origin, starting point, cause and to produce.[41] Adding the character *chuo* 辶 (to walk) to *Yü* 于, the sign of God under heaven, makes up the character *yù* 迂 which signifies "to go astray."[42] *Yúlè* 娱乐 means "fun," and *yúyuè* 愉悦 means "joy, delight." *Hua* 東 means "flower" and China itself."[43] *Yú Huáng Tiān Shàngdì* 天玉上

帝 is the emperor of heaven, who "invented civilization and the compass ⊕."[44] *Yú Hua, Yú Huang,* and *Yao* are all forms of YHWH.

The Power and the Glory

O Lord, our Lord, how majestic is your name in all the earth!
—Psalm 8:1

Aq (center, true mean) is a title of God in Egypt. *Akh* is "glory, splendor." *Aakhu* means "to shine, glorious acts, words of power." *Akhu* is "to become a spirit."[45] The same word in New Zealand means "spirits."[46] *Ak* appears twice in *kokhav* (ok/ok/hav), Hebrew for "star." Enoch (vowed, dedicated), Abimelech (father of a king), Shadrach, Meshach, Gog, Magog, and Habakkuk (embrace) bear this name. *Ikavod* is "the glory has departed" (1 Samuel 4:21). Yoke is the same in Mayan and English. Iugum is "yoked" in Latin. *Yachad* is "union" in Hebrew, and *yoga* is "union" in Sanskrit. The Sanskrit *Ekad* and Hebrew *Echad* both mean "the First, the One."

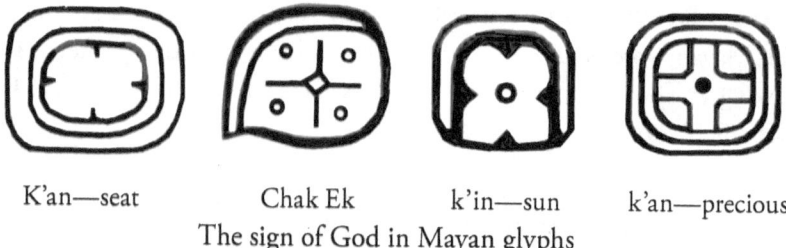

K'an—seat   Chak Ek   k'in—sun   k'an—precious
The sign of God in Mayan glyphs

In Mayan, Xaman Ek is the North Star, Lord of the Night.[47] Ek is "black," because the star of God is now absent. Chak Ek is the morning star. The Mayan Ik is "divine breath, wind, life, spirit,"

On 12 Ik the breath of life was created. The reason it was
called Ik was because there was no death in it.
The Yucatec Mayan *Chilam Balam* trans. Ralph L. Roys

Ikh is also the early word for "soul" in Egyptian. The Huron *Oki* is "divine power." *Waka* is "spirit" in the central Andes. The Mandan four-day flood ceremony is Okipa. Tir na n-Og (Land of Youth) is the Irish heaven. The supreme God of Easter Island is Ko Make Make. The word *light* (Latin *lux,* Greek *leukos,* Germanic *licht)* combines El and Ok.

In Hebrew, *hag* is feast—literally ("to encircle"), *haga* is "to meditate, praise." The names Hagar (flight) and Haggai (Feast of God) bear this name. The Greek word for holy is *hagios.* Hogmanay (Feast of Manay) is New Year's Eve in Scotland.[48] Ig accounts for the once pronounced G in the Anglo-Saxon rooted English words *high, light, fright, might, night, flight,* and *bright,* all having sacred associations. Ag begins the Greek *agape* (love), Sanskrit *ugra* (very strong), and Latin *ignis* (fire). Achrum is the Etruscan name for heaven. Ach is Hebrew for brother, but it is also a German expletive like the Irish oc, the Scottish och and the Native American ugh![49]

Ag/Ok, the prime mover, gives us agent, act, axis, axle, and wagon. Combining *el* and *ak* forms the Babylonian *alaku* and Hebrew *halach* (to go). *Halach* is "Great Lord" in Nahuatl. The root appears in Greek literature in the names Achilles and Agamemnon. Achaia is an ancient name for Greece, and Ogygian is an archaic term for the former age. The d*eluge* is "the Ogygian flood" from Hesiod's Ogygos (king of the gods).[50] Because of the once universal fear of divine judgment, this name begins *achos* (pain, dread) *okpudei* (chilling, horrible), and *agonia* (victims).[51] But *Agnus Dei* (Lamb of God) is the remedy for all.

As the source of all knowledge (Proverbs 3:19), AK/HW is the parent of the Latin *quo, qua, qui,* and the Germanic *who, which, what, when,* and *where.* As the source of all water, His name is found in the following:

English —	*water, wet, wash, wave*	—
Latin —	*aqua*	water
	lava	bathe
Hawaiian —	*awa*	ceremonial drink
Gaelic —	*uisce*	water
Japanese —	*kawa*	river
	ogawa	stream
Scandinavian —	*hav*	sea
Egyptian	*uau*	stream
Hebrew —	*mikvah*	immerse for purification
Arabic —	*wahat, wadi*	water hole, river valley
Narragansett —	*queque*	water
German —	*wasser*	water
Polynesian —	*wai*	water
Russian —	*vada*	water

Chinese — 水 *shui*	water
Swahili — *nywa*	drink

Japanese has many words in the holy tongue. Agameru is "to praise, glorify." The old Japanese verb *aru* is "to appear, to come into being." *Kaga* and *hikari* are "light," and *kagayaku* is "to shine." *Oku* is "to establish," and *iku* 行く and *yuku* are "to go." Agaru is "to rise," *aganai* is "atonement," and *ikiru* is "to live." *Kokoro* is "mind-heart." *Kuwawaru* is "to join," and kokkyu is "breath." Hwh is the root of *wa* (harmony), *niwa* (garden), and *heiwa* (peace). *Kamu-Ya-wi-Mimi-No-Mikoyo* is "Divine-YHWH-Ruler-Lord."

Call upon His Name

For I know the plans I have for you," declares the Lord, "plans to prosper
you and not to harm you, plans to give you hope and a future. Then
you will call on me and come and pray to me, and I will listen to you.
You will seek me and find me when you seek me with all your heart.
—Jeremiah 29:11–13

We greet and take leave of each other in God's name: hello, once *halloa*, hail, *aloha*, and *hau!* The Chinese *hao* 号 is "to call or cry out."[53] Good-bye is a contraction of "God be with ye." His name rings out in the Swedish *skol*, Spanish *salud*, and the English *hale, heal,* and *whole*. Wassail! is the Anglo-Saxon *Was Hál!* (Be whole!). El is the center of *help* (Hebrew *yaal*) as "our help is in the Name of the Lord" (Psalm 124:8a). *Yo-ho tra la la la la la* is an ancient hunting cry as is *halloo* in France, *halali* in Germany, and *tally ho* in Britain.

We can see that our tribal ancestors considered war sacred in the ancient cry "Havoc!" Running *amok* (a Malay word) was a tribe seeing itself as the agent of "divine" justice.[54] The war cry of the ancient Greeks was *"Eleleu, Eleleu Iou Iou."* The Native American war cry *"Hawawawawa"* repeats the Great Spirit's name. *Evocatio* was a Roman rite in which the gods of a besieged city were asked to abandon the inhabitants and side with the attackers in exchange for a cult in Rome.[55]

People who say, "Alas!" (Hebrew *alay*) are calling on God. An *elegy* is a lament for the dead. Wail (Old Norse *vaela*, Gaelic *uaill*) begins and ends with the name of God. Waly! or Walawa! is an expression of profound sadmess in Scotland.[56] We find His name is in the English *woe, sorrow,* and *ill*. From antiquity, mourners at Middle Eastern funerals have chanted, *"Ulululululululululululu!"* A period of calm is a *lull*, and mothers everywhere *lull* their infants to sleep singing *lullabies. Lully, lullay* is the restful opposite of the rousing *Falalalala!*

Word Archaeology

Your name, Lord, endures forever, your renown,
Lord, through all generations.
—Psalm 135:13

There are as many secrets buried in words as in the ground. Knowing the etymology (true meaning) of a word enlightens history. It is apparent that the Chinese *wok,* like *bowl,* once held sacrificial offerings. We can tell by the word that the Scotsman's Haggis is a sacred ritual. We can see that a *hag* was once thought to be a spiritual old woman. Our ancestors incorporated God's name into their tribal centers so they would prosper under His protection (Proverbs 18:10). Place names continue long after their meanings have been forgotten.[57] We hear His name in Acapulco, Avebury, Chihuahua, Ecuador, Guadalajara, Hague, Halawa (Jordan), Havana, Ladakh, Managua, Nicaragua, Okinawa, Oklahoma, Ottawa, Uruguay, Ugarit, Ukraine, Tikal, Tiahuanaco, Quebec, Walla Walla, Warsaw, and Wassau.

All practical arts and sciences imitate God's creative acts. A sacred trust belongs to the farmer, the blacksmith, the weaver, the carpenter, and so on. Words associated with these arts contain one or more of His names. In *agriculture* (Hebrew *agar*—harvest, *ika*— farmer), we have the English words till, soil, sow, grow, harrow, winnow, fallow, furrow, plow (Sanskrit *hala*), acre (Sumerian *iku*), and kernel. Foundry was named for the Founder as well as weld, anvil, alloy, engrave, metal, and tool. In carpentry (Egyptian *aqhu*): ax, log, saw, mill, chisel, drill, level, and nail (Egyptian *agau*).

The Great Fall

How much more, then, will the blood of Christ, who through the eternal
Spirit offered himself unblemished to God, cleanse our consciences
from acts that lead to death, so that we may serve the living God!
—Hebrews 9:14

The word *bless* derives from the Old English word *bledsian* ("to make holy
with blood").[58] The true meaning of "God bless you!" is "God cleanse you
with the blood!" Our forebears knew "it is the blood that makes atonement
for one's life" (Leviticus 17:11). This means innocent blood. The entire ancient
world practiced ritual blood sacrifice to restore their relationship to their
Creator.

Language reflects the painful severance from our heavenly Father and the
downward direction humanity has chosen. *The New World Dictionary* first
defines the English word *fall* as:

Drop, descend
Become detached
To come down suddenly
To be wounded or killed in battle
To come down in ruins, collapse
To lose status, value, dignity
To do wrong, to sin

The late Roger W. Wescott, professor of anthropology and linguistics
at Drew University, and past president of the Linguistics Association of the
United States and Canada, has traced the English word *fall* to a primitive
root *fl/pl*.

English

From Germanic

fell (cut down), flood, flow, flaw (ON)
pluvial, felt, spill, split, fleece, flay, flint,
pelt, freeze, bruise, break, burn, bore,
brimstone, bloat, burst, bolt

From Italic

fail (OF), flail, inflict, flux,
expel, explode, repulse, plume,
press, plague, perforate, inflict,
ferment, fracture, fray

Semitic

Akkadian—palaku (kill), palasu (dig)
Arabic—fala'a (cleave) falaha (plow)
Hebrew—pala (divide), palat (escape)

Our prefixes, he continues, also reflect the great fall.

Latin

de (down) as in depress, decline, defect, deject
dis (apart) as in disintegrate, disappear, disorder, disaster
ex (out) as in expel, exile, exterminate, extinct

Greek

cata (breakdown) as in cataclysm, catastrophe, catabolism
Composite English words—downfall, pitfall, forlorn, forsaken.[59]

"Myth, worldwide, blames humanity for the fall," he states."[60]

European poster using the Tower of Babel to promote unity

The Restoration of the Holy Tongue

So let it be established, that your name may be magnified forever ...
—1 Chronicles 17:24a

All our attempts to establish one universal language have failed because of nationalistic prejudices. Interpreters are needed around the clock, but the same God who inspired Eliezer Ben-Yehuda (1858–1922) to revive ancient Hebrew in Israel after centuries of scattering and persecution has promised to reverse the curse of Babel. "For then will I restore to the people a pure language, that they all may call on the name of the Lord, to serve Him with one accord" (Zephaniah 3:9 NKJV). "And the Lord shall be King over all the earth. In that day it shall be—'The Lord *is* one,' and His name *one*" (Zechariah 14:9 NKJV).

Chapter 7

The River of God

Part 1

The Circle on the Face of the Deep

When he prepared the heavens, I was there: when
he drew a circle upon the face of the deep.
—Proverbs 8:27 (NKJV)

Job, the oldest book in the Bible, contains detail and imagery paralleling the
sacred texts of other nations. In them, Heaven is described as circular. "He
has inscribed a circle on the face of the waters at the boundary between light
and darkness" (Job 26:10 RSV). The Hebrew word *chug* **גוח** (to inscribe a
circle, to encompass, to enclose) used in Job 26:10 is the same word used in
Proverbs 8:27 (above). This circle of glory separated God's holy dwelling from
the rest of space. The word *paradise* derives from the Old Persian *pairidaeza,*
"a round walled enclosure," from *pairi* (around) and *daeza* (a wall).[1]

Ezekiel saw a rainbow halo surrounding the Throne of God: "Like the
appearance of a rainbow in the clouds on a rainy day, so was the radiance
around him. This was the appearance of the likeness of the glory of the
Lord. When I saw it, I fell facedown, and I heard the voice of one speaking"
(Ezekiel 1:28). Similarly, John saw an emerald rainbow encircling the
Throne of God. "There before me was a throne in heaven with someone
sitting on it. A rainbow resembling an emerald encircled the throne"

(Revelation 4:2–3). The rainbow was a reminder to Noah and his family that God was still with them as it was a part of His kingdom.

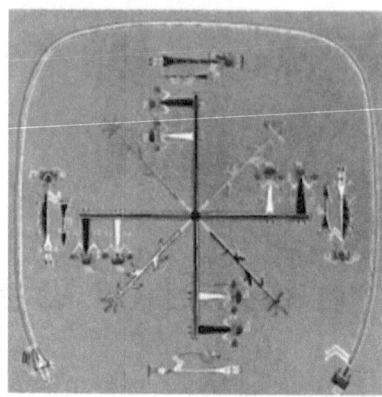

A rainbow surrounds the Navajo night chant ceremony
HB Alexander, "North American Mythology,"
Volume 10 *Mythology of All Races.*

The Crown of Glory

In that day the Lord Almighty will be for a glorious crown,
a beautiful wreath for the remnant of his people.
— Isaiah 28:5

I am the word spoken by the Most High.
I covered the earth like a mist.
I made my home in highest heaven,
my throne on a pillar of cloud.
Alone I walked around the *circle* of the sky
and walked through the ocean beneath the earth.
I ruled over all the earth and the ocean waves,
over every nation, over every people.
(Sirach 24:3–6 GNT)

The circle symbolized perfection and eternity to our ancient ancestors.[2] The Pawnee called it "the Higher Circle of the Father."[3] The Egyptians called it Nebt Seshemu-nifu, the Great Circle of Waters.[4] The circle of God was the boundary between light and darkness, order and chaos. The Lord asks Job,

"Who shut up the sea with doors when it broke forth and issued out of the womb?—When I made the clouds the garment of it, and thick darkness *a swaddling band* for it, and marked for it my appointed boundary and set bars and doors" (Job 38:8–10 AMP). Dante wrote,

> The nature of the universe, which holds the center quiet, and moves all the rest around it, begins here as from its starting-point. And this Heaven has no other than the Divine Mind, wherein is kindled the love that revolves it, and the virtue which it rains down. Light and love enclose it in a *circle* ... this engirdment He alone understands.
> (Dante, *Paradiso* Canto 22 fourteenth century)

Vishnu of India wears the ancient heaven as a crown

The Hebrew term for the crown of glory encircling the Throne of God is Ateret HaKavod. The word is similar to the Old Persian Kvarnah, the "Awesome Royal Glory," which stood for authority and legitimacy, but it fled away.[5] The Sumerians knew it as the Melám. They explained, if the deity leaves his place at the center of the sky, his Melám disappears first.[6] Job refers to this incident, "He hath stripped me of my glory, and taken the crown from my head" (Job 19:9). Like Vishnu, the Vedic Indra also "wears heaven as his crown."[7]

Tertullian quotes Pherecydes of Syros (sixth century BC), who taught the immortality of the soul, as saying that the Greek god Kronos was "the first before all others that ever wore a crown."[8] The high priest of Israel wore a golden crown engraved with "Holiness to the Lord" (Exodus 29:6, 39:30).

According to the Talmud, a circle was drawn in anointing kings.[9] The Celts memorialized the crown of glory in their metal torcs worn around their necks for divine protection.

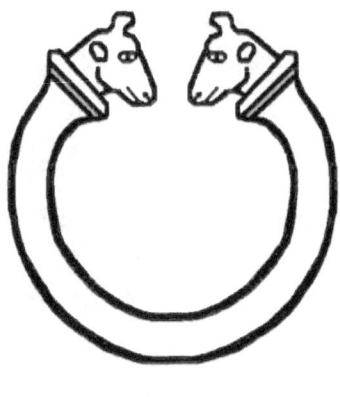

Celtic torc

The Celestial River Jubal

Make a joyful shout to the Lord, all you lands! Serve the Lord
with gladness; come before His presence with singing.
—Psalm 100:1, 2 (NKJV)

The Hebrew word for "the heavens" in Genesis 1:1 is *shamayim,* a compound word meaning "there-waters." Josephus mentions a circular river that is parted into four streams encircling Eden.

> Now the Garden was watered by one river, which ran round
> about the whole earth, and was parted into four parts.
> (*Antiquities,* book 1, chapter 1)

This is the same river as Psalm 46, which is clearly in the heavens. "There is a river, the streams whereof shall make glad the city of God, the holy place of the tabernacles of the most high. God is in the midst of her; she shall not be moved" (Psalm 46:4, 5 KJV). We also find it in the writings of the early Christian fathers:

And he set me upon the river whose source springs up in the Circle of Heaven, and it is this river, which encircles the whole earth. And he said unto me; this river is Ocean.
(*The Apocryphal Gospels, Acts and Revelations*, Ante-Nicene Christian Library)

No matter how it was interpreted by those who had not seen it, this "river of ocean" was never on this planet. The prophet Daniel calls it the Great Sea, "In my vision at night I looked, and there before me were the four winds of heaven churning up the great sea" (Daniel 7:2). The river of Psalm 46:4 is the same river as Genesis 2:10–14. Here the "streams that make glad the City of God" are named, but the circular river is not.

In the Midrash Konen, the circular river is called Yuval (river, stream), Yubal in Aramaic.[10] This word, also spelled Jubal, is connected everywhere to rivers, light, shouting, praising, and singing. Yebul is a chant of praise in Buddhism. It begins with the ancient character *Ya*卍. The Chinese word *yibu* 医卜 means "medicine." The Irish hero Cuchulainn owned a chariot driven by Iubar,[11] the Latin equivalent of yubal. The following words are related:

Hebrew/Aramaic

yuval/yubal—river, stream

yobel—shout, trumpet blast

yabal—lead, conduct

yebul—growth, increase

Japanese

yobu—to call out, to nail

yubiwa—ring

Latin

iubar—light, radiance, star

iubilatio—shout, rejoice

iubeo—order, command

Babylonian

Hubur—the perfect river

Egyptian (R = L)

ibr—river, brilliance, splendor

Since it was from within Yubal that the Creator governed the universe, the word *government* stems from the related Latin word *guberno* (to steer a ship). *Gubernaculum* (steering oar or helm) reminds us of Plato's heavenly helmsman,

In the fullness of time, when the change was to take place, and the earth-born race had all perished ... The Pilot of the

Universe let the helm go, and retired. (*"The Statesman"* trans. Benjamin Jowett)

Okeanos

But if I were you, I would appeal to God; I would lay
my cause before him. He performs wonders that cannot
be fathomed, miracles that cannot be counted.
—Job 5:8, 9

The Greek version of Yubal was Okeanos. Hesiod called it "the current, which turns in on itself" and "the Perfect River."[12] Plato wrote,

There are four principal (rivers), of which the greatest and outermost is that called *Okeanos,* which flows round the earth in a circle. (*The Dialogues of Plato,* "Phaedo" trans. Benjamin Jowett)

Homer called it "the place from whom the gods are sprung" (*Iliad* book 14). Okeanos encircled two crossed rivers: the River of Flaming Fire and the River of Lamentation.[13] The word *ocean* derives from Okeanos, not the other way around. One of the most beautiful translations of the Orphic Hymn to Okeanos is by Thomas Taylor.

Okeanos whose nature ever flows,
From whom at first both gods and men arose;
Sire incorruptible, whose waves surround,
And earth's all-terminating circle bound:
Hence every river, hence the spreading sea,
And earth's pure bubbling fountains spring from thee.
Hear, mighty sire, for boundless bliss is thine,
Greatest cathartic of the powers divine:
Earth's friendly limit, *fountain of the pole,*
Whose waves wide spreading and circumfluent roll.
Approach benevolent, with placid mind, and be forever to
thy mystics kind. —To Okeanos

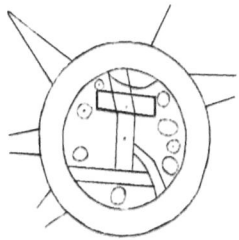 

Babylon
600 BC

Anaximander
c. 610-546 BC

Above, we see two old-world maps showing the earth encircled by Okeanos. Greek historian Herodotus (fifth century BC) ridicules this concept:

> I laugh to see how many have before now drawn maps of the world, not one of them reasonably; for they draw the world as round as if fashioned by compasses, encircled by the Okeanos river, and Asia and Europe of a like extent. (Herodotus, *Histories* 4 trans. AD Godley)

Solomon's Sea and the Abyss

And he made the Sea of cast bronze, ten cubits from one brim
to the other; it was completely round. Its height was five cubits,
and a line of thirty cubits measured its circumference.
—1 Kings 7:23 (NKJV)

Solomon constructed a bronze laver in front of the temple in Jerusalem. Supported by twelve bulls in the form of a *tav*, it was thirty feet long, thirty feet wide, fifteen feet high, and held 17,500 gallons. "The Sea stood on twelve bulls, three facing north, three facing west, three facing south and three facing east. The Sea rested on top of them, and their hindquarters were toward the center. It was a handbreadth in thickness, and its rim was like the rim of a cup, like a lily blossom. It held three thousand baths" (2 Chronicles 4:4, 5). Solomon's sea enabled the priests to purify their hands. It was later broken up and the pieces carried off to Babylon (2 Kings 25:13).

Solomon's Sea where the priests washed their hands.
George A. Barton, *Archaeology and the Bible* (1925).

The ancients saw Jubal as the divine source of all earthly rivers and streams, making all of them sacred. Dipping seven times in a river symbolized God's rejuvenating power (2 Kings 5:10). In Mesopotamia, the cuneiform sign for God ✝ began the word river. In hymns, the river was called Mercy.[14] The name India derives from *sindhu*, Sanskrit for river, flood, ocean, and symbolic of the number four. Every river in India is seen as part of the heavenly Ganga (Ganges).[15] Its waters make holy seven generations:

> He that reciteth the name of Ganga is purified; while he that beholdeth her, receiveth prosperity; while he that bathes in her and drinks of her waters, sanctifieth seven generations of his race up and down. (*The Mahabharata* trans. Protap Chandra Roy)

The ancients offered sacrifices when crossing rivers. Hesiod wrote:

> Whoever crosses a river with hands unwashed of wickedness, the gods are angry with him and bring trouble upon him afterwards. (*Theogany* trans. Hugh G. Evelyn-White)

The vestal virgins threw straw men *(argei)* from a sacred bridge into the Tiber on the ides of May. According to Dionysus of Halicarnassus, these replaced the twenty-four old men that were formerly cast in as a sacrifice to Saturn.[16] The Trojans sacrificed to the Scamander, the Cimbri to the Rhone, and the Chinese to the Yellow River.[17] Communal cleansing in rivers was conducted at the New Year festivals. People tore their clothes, and sins were symbolically cast in (Joel 2:13). The tradition of spring-cleaning and the wearing of new clothes at Easter or White Sunday are remnants of this once universal purification.[18]

The Japanese festival of Oharai (Great Purification) takes place in Shinto temples on December 31 and June 30. It involves a confession of sins, which are transferred to rice stalks, rags, and animal hides and then cast into rivers. Everyone writes his name and birthday on a life-size paper doll, which is breathed on and rubbed against the body. These are tied into bundles and thrown into streams. The priest waves a broom, symbolizing the sweeping away of sins, and the penitent's clothing is destroyed.[19]

Asshur, god of Assyria, seated within the ring.
D. Mackenzie, *Myths of Babylonia and Assyria* (1915).

After Babel, Jubal had many names. The Welsh called it Arianrhod, the polar home of the gods; the Scandinavians called it Ginnungagap, the great gulf surrounding Valhalla.[20] As the barrier separating God's kingdom from chaos, being "thrown into the abyss" was to be placed in the purifying stream far from the center where God lives (Luke 8:31, Revelation 20:3). The Egyptian army who perished in the Red Sea was a type of the wicked perishing in the abyss. Israel's bondage was a "type" of humanity's bondage in microcosm.

In Mesopotamia, Jubal was called Abzu (nether sea), the watery abyss, and progenitor of the gods. A circle of rope (tarkullu) represented it. It was also called Mother Hubur, the River of Death.[21] Abzu was also a tank of holy water in the temple courtyard.[22] The living cannot cross these waters. The hero Gilgamesh wished to enter the Land of the Crossing Rivers, but the gatekeeper discouraged him,

> Gilgamesh, there has never been a crossing (here), and no one since eternal days has ever crossed the sea. Shamash, the hero, crosses it; but besides Shamash who can cross it? Difficult is the crossing, and extremely dangerous the way, and closed are the Waters of Death, which bolt its entrance. (*Epic of Gilgamesh*, Tablet X[23])

Seal Cylinders of Western Asia (William Hayes Ward 1910)

The Egyptian Ring of Eternity

> He set the springs of ocean in their place and prescribed limits for the sea.
> —Proverbs 8:29

The common Egyptian name for the Celestial River is Nu (ring, circle). It is "the heavenly source of the Nile.[24] Nu enclosed the Otherworld and the throne of Ra ☉.[25] In the texts, Nu is called "the Cosmic Ocean, the Celestial Nile, the Great Circle, the Ring of Eternity." Nu is the basic principle of cosmology common to all accounts of origins.[26] The feminine form is Nu-t, which also means "town" and "city." Nu-t Neter is "the City of God."[27] Nu-t is the source of the Coptic word for "abyss or deep."[28] In some texts, it is called "a great flood."[29]

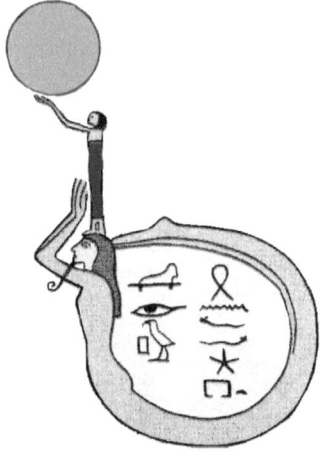

Nu-t, the earliest Egyptian personification of the
Celestial River (Max Müller 1918)

A common emblem of Nu and eternity was the *shen* ◯. As in Mesopotamia, its hieroglyph was a circle of rope seen in the hands of gods and kings symbolizing their right to rule. The priests placed the hieroglyphic names of gods and kings within the *shen*. To fit longer names, the shen was stretched out, becoming the royal cartouche ⬭.[30] The feminine form of the word is *shenit*. It represented the "officials of the court of Osiris."[31] We use this word today, inheriting it from the Romans, who borrowed it for their senate, the council of Rome. *Sen* means "old" or "a council of elders."

The Eye of God

The Eye of the Lord is upon them that fear him,
upon them that hope in His mercy.
—Psalm 33:18

When the Calumet Indians refer to "the Eye of the Great Spirit," the Polynesians to "the Eye of Tané," or the Mayans to "the Great Eye," they are using the terminology of the previous age. Hesiod wrote, "The eye of Zeus seeing all, and understanding all, beholds these things."[32] The throne of God within its crown of glory resembled a great eye in the sky.

In the Babylonian Talmud, Rabbi Yose made this statement: They waxed haughty only on account of the covetousness of the *eyeball, which is like water,* as it is written, and they took them wives from all which they chose. Therefore He punished them by water, which is like the eyeball (Galgal HaAyin) as it is written, all the fountains of the great deep were broken up, and the windows of heaven were opened. (Tractate Sanhedrin 108a trans. Rabbi Dr. I. Epstein)

The sixteenth letter of the Hebrew alphabet is *ayin* ע. As a word, it means both "eye" and "fountain," because the Dayspring on High, the Fountain of Life, is within the Eye of God.

The Egyptians associated the Eye of God with the Creative Word.[33] The eye (Udjat) is the most common Egyptian symbol, but it is very strange to us.[34] The word derives from *utch* (to order, command, send out), which is related to *utcha* (to be healthy, sound, safe, strong, whole).[35] As long as the Eye of God was in place, humanity was safe. In the texts, the eye in the primordial ocean is called "the flood" and "father of the gods."[36]

The importance of the Eye of God is seen in the oldest hymns.

Praise be to thee, O Ra, exalted *Sekhem* (Power), aged one of the pupil of the *Udjat* (Eye), thou makest full thy splendor.

Praise be to thee, O Ra, exalted *Sekhem Bua-tep,* governor of his Eye; thou sendest forth light into the hidden place.

Praise be to thee, O Ra, exalted *Sekhem*; when thou fillest thine Eye and speakest to the pupil thereof, divine dead bodies shed tears.
(The Seventy-Five Praises of Ra)[37]

According to Egyptologist Rundle-Clark, the eye is the key to the religion. The literature of all periods, including the oldest—the Pyramid Texts—constantly alludes to its loss and the bringing back of it. Its calamitous disappearance plunged the country into darkness, gloom, sterility, fear, and lifelessness.[38] Originally, the High God had only *one* eye, which was "separate from the god himself." It was a kind of "original sun," which he sent forth into the primeval abyss.[39]

The Lost Eye
Max Müller, "Egyptian Mythology," *Mythology of All Races,* Vol. 13 (1918).

The Egyptians had several explanations for the loss of the eye. In one, a daughter of Ra is sent out by her father to punish human beings. In another, there is a fight between two gods that blinds the eye.[40] The eye belonged to all the creator gods of all the nomes of Egypt. A Nile hymn mentions "the flood of the Eye of *Atum,*" a title of Ra.[41] The original Divine Eye ☉ is the hieroglyph for both Ra and day. The Eye of Horus, the morning star, is constantly mentioned in the Pyramid Texts. The original form of the Greek letter *omicron* was ☉, but the "pupil" was later lost. It is the source of the letter O (once resembling a human eye) in the early Semitic alphabets.[42]

The word *eye* is feminine in both Egyptian and Hebrew and called "the daughter" in both languages. Although most of our Bibles have "apple" or "pupil," the Hebrew of the psalmist is this: "Keep me as the daughter *(bat ayin)* of your eye; hide me in the shadow of your wings from the wicked who assail me from my mortal enemies who surround me" (Psalm 17:8) and "For this is what the Lord Almighty says: 'After the glorious one has sent me against the nations that have plundered you—for whoever touches you touches the (daughter) of his eye'" (Zechariah 2:8).

Since the Lord did all His work within His eye, the ancient root *op*, which means, "to produce abundantly," is the source of our eye words—optical, myopia, and ophthalmology. Horus of Egypt shines out of his throne within his eye:

> I am in the sun's eye when it closes, and I live by the strength thereof. I come forth and I shine; I enter in and I come to life. I am in the sun's eye, my seat is on my throne, and I sit

within the eye. I am Horus who passes through millions of years. I have governed my throne and I rule it by the words of my mouth. (EAW Budge, *The Book of the Dead,* chapter 42)

Op also means best, abundant, hence the Latin *opis* (plenty) and *optimus* as in optimistic, opulent, and copious. The most important eye word is *open.*

The Divine Eye at the pole is part of the heritage of every tribe. Japanese texts mention Ama-no Ma-hitotsu (One-Eye of heaven), Miyau Ken (divine, mysterious eye), and Ama-no-ma-hito-tsu-no-mikoto (His Augustness Heavenly-One-Eye).[43] The eye was found on oracle bones in ancient China. The original character *jih* 日 (sun and day) was ☉.[44] The Sumerians called it "the Eye of Life and Death." Enki (Akkadian Ea) is Bel-ini-elli (Lord of the far-seeing eye). At the Flood, Enki's eye convulsed the Kur (netherworld).[45]

Changing the original glyph to resemble a human eye added to the confusion. It produced countless tales of a one-eyed god like the Norse Odin, who wears an eye patch. Avalokiteśvara was kept in the door hinge of the eye. Krishna had an all-seeing eye, and Mitra had an unwinking eye that sees all. Some deities are represented with a "third eye" in the *center* of the forehead. The Persian Ahura Mazda had an all-piercing eye. The Irish hero Cuchulainn had a monstrous eye that blazed forth like the sun.

Eventually, the eye became just a bright star.[46] The Flood caused it to be viewed as dangerous. The Irish Balor had a "baleful eye" with seven coverings. The Greek Cyclops (wheel-eye) was a one-eyed cannibal giant blinded by Odysseus and later Sinbad. Eye amulets and talismans are universally seen as providing protection. The all-seeing eye remains a powerful symbol everywhere.

The Egyptian word *netr* (gods, stars) also means eye. The related Sanskrit word *netra* means "eye, pipe-tube, veil, root of a tree, carriage, and river."[47] Some of the founding fathers of the USA were Freemasons. The pyramid with an eye on it is a popular symbol of God's architecture. We see it on the one-dollar bill and the great seal as a target to aim for. In like manner, Joseph Campbell rightly compared the bull's-eye to the sun door of the Otherworld.[48]

Eye amulet (Frederick T. Elworthy)

Celestial Geography

Samuel said to the people, "Come, let us go to
Gilgal, and there renew the kingship."
—1 Samuel 11:14

Since all human beings once lived in a united territory directly under the Kingdom of God, when they divided after Babel, each tribe took the same universal history with it. Wherever they settled, they tried to recreate His holy ground. All early settlements began with a circle drawn from a central point, symbolically repeating the creation. A circular trench inscribed with a plow marked off the boundary between sacred space and chaotic space, the habitation of evil forces.[49]

The ground circle is the most ancient place of worship.[50] The word *church* derives from the Greek *kirke* (circle). Joshua transformed a local ground circle into a stone circle after Israel crossed the Jordan. "And Joshua set up at Gilgal (circle) the twelve stones they had taken out of the Jordan" (Joshua 4:20). Gilgal was a place for assembly, judgment, and coronations (1 Samuel 7:16, 11:15a). Thousands of ring ditches with human and animal bones were later set in stone. Stonehenge is the best known but not the largest of these. Avebury, with its thirty-feet-deep ditch, is a megalithic complex encompassing the entire village of Wiltshire, over twenty-eight acres![51] See Rujm el-Hiri, Israel's Stonehenge on the Golan Heights (3000 BC).

Destruction of the tribal circle was a reversion to chaos. Within the circle is life, health, and civilization, and without are the barbarians. Walls, moats, and ramparts were consecrated as a spiritual defense long before they were a military one.[52] In every tribe, we find the remnants of this sacred geography. Tribal lands were subdivided into four regions imitating the King of Heaven's "Four Courts of Praise." These were arranged around God's central "fifth" region. Each tribe saw its own city as the only true replica of the City of God.[53]

While he was second in command in Egypt, Joseph gave Pharaoh the fifth part of the harvest because, as heir of Ra, it was rightfully his. "And Joseph made it a law over the land of Egypt to this day, that Pharaoh should have the fifth part" (Genesis 47:26). One's abode is still called his "quarters," and the quartermaster's domain is his headquarters. The tribal chief or king governed the four quarters from the central fifth region, which became the capital, from the Latin *caput* (head). Earthly kings asked for a fifth, but God only asked His people for a tenth (tithe).

The Assyro-Babylonians divided their land into Akkad (south), Elam (east), Subartu (north), and Amurru (west). Rulers, such as Sargon of Akkad, were addressed as "King of the Four Quarters."[54] Cyrus also took this title: "I am Cyrus, King of the Universe, the great king, the mighty king, king of Babylon, king of Sumer and Akkad, king of the four quarters of the world" (The Cyrus Cylinder[55]).

The center and four quarters formed the original territorial unit of Ireland. Provinces Munster, Leinster, Ulster, and Connaught formed a pentarchy with the central Midhe (Meath). The Gaelic word *cóigedh* (province) literally means "a fifth."[56] Within Midhe was Tara, "the Navel of Ireland," where the people held sacred assemblies and inaugurated kings. The same is true of the Celts of Gaul (France). Julius Caesar wrote that the Druids of Carnutes gathered at a fixed time of the year in a sacred assembly at the heart of Gaul.[57]

The Chinese celestial emperor Shangdi上帝 ruled the four quarters from his royal domain in the center. There are five sacred colors, five metals, five tastes, and five elements. The spirits of the Five Elements are the Five Ancients. The pentatonic scale is the foundation of Chinese music. The earthly pentad symbolizes a universe complete and in order maintained by a righteous Son of Heaven.[58]

The Aztecs laid out their cities as a *quincunx*, four axial roads leading out from the center to the four directions.[59] The word *quincunx*, derives from the root *penque* (five). Also from penque derives the Germanic *fingers, fifth*, the Latin *quinque,* Greek *penta,* and Sanskrit *panta.* Punch was originally a sacred drink with five ingredients.[60] The Mayans also had four directional towns arranged around a central "fifth," a plan seen at many sites.[61]

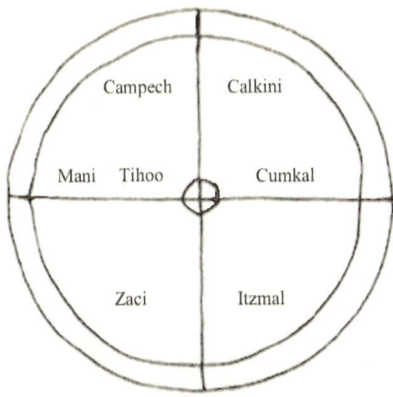

Ancient divisions of the Yucatan, *Book of Chilam Balam*

The Recreation of Heaven on Earth

There shall not be found among you anyone who makes
his son or daughter pass through the fire, or who uses
divination, or is a soothsayer, or an *augur,* or a sorcerer.
—Deuteronomy 18:10 AMP)

All tribal life revolved around the altar and the temple, a word that derives from the Latin *templum* (to cut off) a consecrated place. The templum was circumscribed and *inaugurated. Aug* (to increase) is an ancient Latin word from which *augus* (full of divine force) derives.[62] The entire ancient world once practiced this form of divination. Only the Romans had a complete system with fixed rules, and only patricians could be augurs elected for life.[63] The title Augustus means "holy, or consecrated by augurs." Cicero held the office of state augur.

Augury was a rite carried out in secret at midnight. To lay out a camp or settlement, the augur held his staff (*lituus*) resembling a shepherd's crook, in his right hand. He looked up into the North Polar heaven along the axial

Cardo Caeli and marked out a cross in the sky. *Cardo* (pole, hinge) is the root of *cardinal.* The mound on which he stood was the Templum.[64] On the ground he first drew the Cardo Maximus (north-south line) followed by the Decumanus (east-west line) at right angles to it. A bull and cow yoked together plowed a circular trench around the site. This became the sacred boundary or *mundus.* Where a gate was to be, the plow was carried across. Every Roman camp had an *augurale* for this purpose.[65]

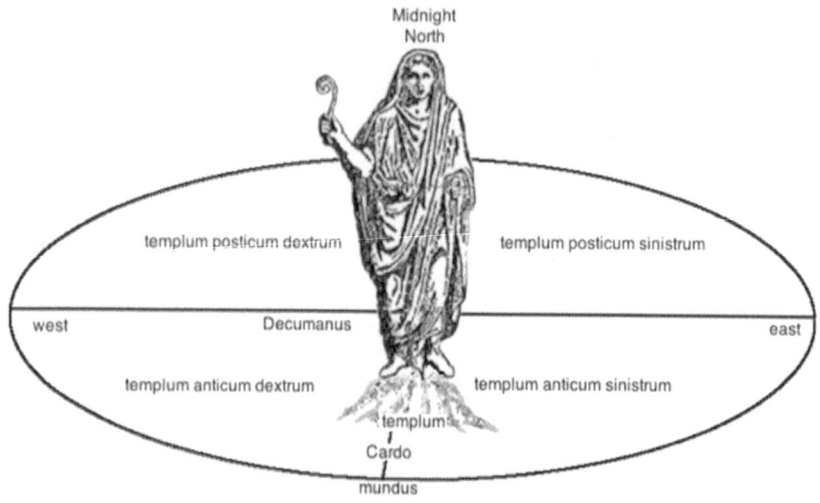

Roman augury *Nordisk familjebok*

When the augur had defined a Templum, he pitched his tent— *Tabernaculum capere.* The Templum was later enclosed with planks, and curtains were attached to four posts fixed in the ground. It had only one door (exitus).[66] A font was placed in front for purification. Within Rome, no tent was necessary, for a place on the summit of the Capitoline called Auguraculum had been consecrated for all time.[67] Roman cities were the *urbes,* from *orbis* (round). The verb *urvare* means, "to ritually plow around the site of a proposed city." The similar Greek *ouron* is "a boundary marked by a plow." The Hebrew counterpart is *chug* (to inscribe a circle) and *habar* (to divide the heavens). This act of drawing a cross within a circle on earth ties in with the name *Heber* (crossing) from which the word Hebrew derives.

Native Americans have a similar traditional respect for their camp circle in which they celebrate the worship of the Great Spirit. In their pipe ceremonies, they blow smoke to the Four Corners, the Middle Point, to heaven and to earth. Their sacred wheel—*Ho hoti*—is seen as the hoop that represents the whole nation.[68] As Black Elk lamented, "The nation's hoop is broken now and there is no more center."[69]

We Shall Gather at the River

And he showed me a pure river of water of life, clear as crystal,
proceeding out of the throne of God and of the Lamb.
—Revelation 22:1

The circle of God was the model for the wedding ring, the laurel and Christmas wreath, and all our round cakes (bagels, doughnuts, etc.). It even inspired a cycle of operas, Wagner's *Ring of the Niebelungen*. In *Die Walküre*, Wotan encircles the sleeping Brünhilde by a ring of fire. Odin (German *Wotan*, Anglo-Saxon *Woden*) had his home in Asgard surrounded by rings of oceans and mountains.[70] After the Flood, earth's ocean and the Milky Way inherited the lore of the River of God, although they are not visibly circular.

Part 2

The Sevenfold Light

Moreover the light of the moon will be as the light of the
sun, and the light of the sun will be sevenfold, as the light
of seven days, in the day that the Lord binds up the bruise
of His people. And heals the stroke of their wound.
—Isaiah 30:26 (NKJV)

When Isaiah prophesied a sevenfold increase in the light of the sun, he was not speaking of the present sun, for if that were to happen, all life on earth would be extinguished. What then does he mean by "the light of the sun will be sevenfold, as the light of seven days"? If asked, our ancient ancestors would say, "It was the light of the Creation." According to the rabbis, this sevenfold light remained visible until the Flood, and all the light now left in the world is only one seventh of this holy light.[1] It is the primary reason the Creator is associated with the number seven. In Jewish tradition, seven attributes serve the Throne of God, wisdom, justice, law, grace, mercy, truth, and peace.

The number *seven* is found 735 times in the Bible, and *seventh* is mentioned 119 times. Seven expresses perfection and completion. A prime number with no divisors, it is made up of three, the number of the Triune God, plus four, the number of the earth. Seven is also the number of life, and the Book of Life is mentioned seven times in the Bible. Life operates in cycles of sevens. Most gestations, including human beings, are multiples of seven.[2] There are seven notes in the diatonic scale and seven colors of the rainbow. The ancients recognized seven liberal arts: grammar, logic, rhetoric, arithmetic, geometry, music, and astronomy.

Noah brought seven clean animals into the ark, and the seventh time his name is mentioned, he is called "perfect in his generations" (Genesis 6:9). There are seven blessings of Abraham, and God instituted seven holy convocations in Leviticus:

> Pesach (Passover)
> Hag HaMatzot (Unleavened Bread)
> Yom HaBikkurim (Firstfruits)
> Shavuot (Pentecost)
> Yom T'ruah (Day of Blowing [Trumpets])
> Yom Kippur (Day of Atonement)
> Sukkot (Tabernacles)

Jesus said to forgive "seventy times seven" (Matthew 18:22). Seven petitions are in the Lord's Prayer, and Jesus spoke seven words from the cross.

The book of Revelation is built around a system of fifty-four sevens: Jesus gave seven letters to His seven congregations. There are seven archangels, seven seals, seven trumpets, seven stars, seven lamps, seven thunders, seven mountains, seven kings, and seven blessings:

> Blessed is he that reads this prophecy. 1:3
> Blessed are the dead who die in the Lord. 14:13
> Blessed is he that watches (for the Lord's coming). 16:15
> Blessed are those called to the Marriage Supper of the Lamb. 19:9
> Blessed is he that has a part in the First Resurrection. 20:6

Blessed is he that keeps the words of this book. 22:7
Blessed are they that wash their robes. 22:14

The Sabbath, a Sevenfold Delight

Call the Sabbath a delight, the holy day of the Lord honorable ...
—Isaiah 58:13a (NKJV)

The Hebrew word Shabbat (Sabbath) derives from the Semitic root *sh-b-t* meaning "to stop or rest." In Babylon, the middle of the month was called by a related word *shapattu*.[3] Shabbat also means "to seven oneself." The Sabbath commemorates the day God rested from His creative work. The Sabbath existed before the Law. Its observance indicated the holiness of the people (Numbers 15:32–36). The Sabbath is a joyous family day set apart for rest and refreshment.

The seventh month, Tishri (Sept./Oct.), is the sabbatical month and contains the most holy days. Every seventh year is a sabbatical year for the land to rest. God provided enough in the sixth year to carry through the seventh. Every seventh sabbatical year (forty-nine years) is followed by a Jubilee in the fiftieth year, a period of redemption. There is repentance, cleansing from sin, reconsecration, and release from bondage. It begins with the blowing of the *shofar*. All slaves are set free, all debts canceled, and all lands previously sold were returned to their original owners.

The sevenfold walled castle motif by Botticelli (WR Lethaby)

The Jubilee is named for Jubal, the Sevenfold River of God. This rainbow river is remembered as the seven levels of heaven - Dok, Shechakim, Zevul, Maon, Machon, Aravot, and Rakia.[4] The Jubilee is a foretaste of the Grand Sabbath Jubilee, the seventh millennial one thousand years of Messiah's reign.

On this long-awaited day, nature will be released from the curse placed on it at the fall. There will be a renewal of the entire cosmic order. "All the host of heaven shall be dissolved, and the heavens shall be rolled up like a scroll; all their host shall fall down as the leaf falls from the vine, and as fruit falling from a fig tree" (Isaiah 34:4 NKJV; see also Revelation 6:14).

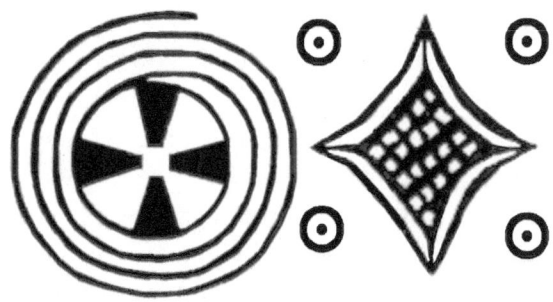

Paradise motifs on Philistine pottery (ES)

The Sabbath Marriage

> I will betroth you to me forever; I will betroth you in
> righteousness and justice, in love and compassion. I will betroth
> you in faithfulness, and you will acknowledge the Lord.
> —Hosea 2:19, 20

The Sabbath is closely associated with the sacrament of marriage. There are seven verses in the Bible, which state, "When a man takes a wife." Weddings often take place on Sabbath eve, as it is the weekly celebration of the *hieros gamos* (marriage of heaven and earth). The Sabbath is called Kallah (the bride) as it is a symbolic wedding between God and His people. Special Sabbath garments are put on to greet the King of the Universe[6] (Revelation 19:13). The Grand Sabbath Jubilee is known as the Sabbath of the Bride when God rejoices over His beloved. "As the bridegroom rejoices over the bride, so shall your God rejoice over you" (Isaiah 62:5b AMP).

The Orthodox marriage celebrates the Sabbath. It is seven days long (Genesis 29:27, 28) and seven benedictions are said. When the bride reaches the canopy (Huppah) held up by four poles, she circles the groom seven times, a tradition based on the verse "For the Lord has created a new thing in the earth—a woman shall encompass a man" (Jeremiah 31:22b NKJV). In the Mitzvah Tantz, the guests revolve around the bride seven times. A group of males lift the chair of the bride over their heads and whirl her around.[7] The Sabbath is the traditional day for lovemaking between husband and wife, for it is believed that the Shekinah (Divine Presence) is with the couple at this time. The rabbis saw that the Hebrew for man is *ish* and his wife is *isha*—similar but different. The letters making the difference are *yod* and *heh*, one of the names of God (Yah).[8]

Sevenfold circumambulation is retained in other marriage traditions. During the Gypsy marriage ceremony, the bride and groom circle the fire seven times to show they are bound together.[9] The Celtic bride is taken seven times around her future home, and the glass goes around the company the same way.[10] A marriage is irrevocable when the Hindu bride and groom take seven steps together.[11] In all of these, the couple is treated as a king and queen because of whom they represent.

The Highest Heaven

To the Lord your God belong the heavens, even the highest heavens, the earth, and everything in it.
—Deuteronomy 10:14

The rabbis teach that within the seventh heaven there is an eighth "fixed" sphere called the heart.[12] The highest heaven, where God dwells (1 Kings 8:27), is simultaneously the center, the highest, the first, and the last. It represents the divine principle, "The last will be first, and the first will be last" (Matthew 20:16). The number eight (Greek *okto*, German *acht*, Japanese *hachi*) is the number of rebirth. There are eight appointed times: the Sabbath and the seven feasts (Leviticus 23). Jesus arose from the dead on the eighth day, Sunday, also the first. Eight people, apart from the Lord, were also raised. Eight people survived the Flood. Circumcision, the sign of the covenant, is performed on the eighth day (Genesis 17:11, 12).

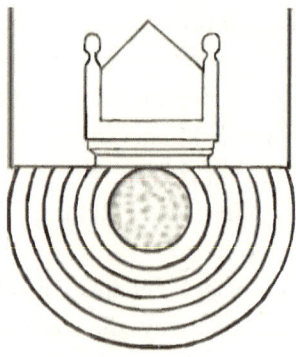

The Throne of Majesty with Seven Steps, thirteenth-
century manuscript—WR Lethaby.
Inscription: "Seven steps in fashion of a hollow vault."

The New Year Festivals

There remains, then, a Sabbath-rest for the people of God.
—Hebrews 4:9

The Lord told Joshua to march the people around Jericho while seven priests blew seven trumpets before the ark. On the seventh day, they marched around seven times while blowing trumpets (Joshua 6:4). "By faith the walls of Jericho fell after they had been encircled for seven days" (Hebrews 11:30 NKJV). This once universal rite symbolized a new beginning, which required a ritual reenthronement of the Creator as king of the universe. This annual restoring of the divine order protected the entire tribe. Many also observed a ritual marriage between the king and the land.

Symbol of Diwali (ES)

All New Year festivals have common elements. All were festivals of light beginning at sunset (Genesis 1:5b). These include the Roman Saturnalia, the Greek Kronalia, the Persian Nauroz, the Hindu Diwali, the Chinese Dragon-boat festival, the Japanese Feast of Lanterns, the Anglo-Saxon Yule, the German Fasching, the Assyrian Sacea, and Babylonian Akitu. Originally seven to twelve days long, there was purification, personal contemplation of one's fate on the Day of Judgment, confession of sins, expulsion of evil, and extinguishing and rekindling the sacred fire. Lots were cast to determine the fate of the tribe for the coming year.[13]

All had processions around the community to drive out the evil of the old year and renew life. Participants (*guisers*) wore masks of animals, ancestors, and demonic figures. It was believed that ancestral spirits returned to the earth on the New Year to wreak havoc. To scare the angry ones away, the people disguised themselves, shouted, blew horns, and beat drums, creating pandemonium.[14] Mock battles were fought between the servants of the tribal god and the devil, the spirit of disorder, often represented by a dragon, and his army. Bands of children stopped at houses, asking for offerings.

There was usually dancing around a decorated tree or pole representing the tree of life. There was gift giving (colored eggs were popular), drinking, and feasting on special sweet foods, recalling the abundance of the golden age. There was a reversion to chaos symbolizing the communal license at the end of that period. There was an inversion of ranks with prisoners released and servants changing places with their masters. A mock king was chosen and anointed and rewarded for a time with all the privileges of royalty then dispatched at the end of the feast, replacing the tribal king originally killed.

Many tribes have more than one New Year. Dates multiplied due to the adoption of different calendars among conquered peoples. The Roman year began on March 1, and those Rome ruled celebrated their New Year at the old time as well as the new. Mardi Gras and Carnival are spring Saturnalias with many ritual elements. The holiday was formerly a pre-Lenten Shrovetide, a confession of sins before the fast. Unable to stamp it out, the church "Christianized" it. Instead of forty days of mourning for what was lost in the Flood, there was now a forty-day fast honoring Christ. Originally, there was a black bean hidden inside a cake, representing a sacrificial victim. This is now a tiny doll representing the Christ child. A mock king rides the main streets as the star of the feast.

The Celts observed festivals at the cardinal points of the year: Imbolc (February 1), Beltane (May 1), Lammas (August 1), and Samhain (November

1). All begin at sunset. Of these, Beltane (fire of Bel) and Samhain (assembly) were symbolic returns to the primeval time, the dissolution of the established order and recreation of the year.[15] The church renamed Samhain Eve (October 31) "All Hallows' Eve," which became Halloween. November 1 was renamed "All Saints Day." This substituted honoring dead saints for dead ancestors. These four cross-quarter feasts, plus the solstices and equinoxes, resulted in an eightfold division of the year.

Most New Year feasts are no longer recognizable as celebrations of the creation, repentance, and rededication to God. The seven times around torchlight processions to sanctify the village have been replaced with gaudy parades through the main thoroughfares. All are still elaborate masquerades and symbolic returns to the prosperity, equality, and license at the end of the antediluvian era. Social norms are inverted, and all restraint is abandoned.

The Jewish New Year Feasts

Therefore do not let anyone judge you by what you eat or
drink, or with regard to a religious festival, a New Moon
celebration, or a Sabbath day. These are a shadow of the things
that were to come; the reality, however, is found in Christ.
—Colossians 2:16–17

The rabbis taught four judgments and four new years.[16] Passover begins on the fourteenth of Nisan (March/April), the first month of the religious year (Exodus 12:1, 2). The people search for and burn *chometz* (any fermented mixture of flour and water, a symbol of sin). The Feast of Unleavened Bread is observed on the fifteenth day (Deuteronomy 16:3). For seven days, the people eat matzot, a symbol of Christ, who is the Bread of Life (John 6:35). The Feast of Firstfruits is celebrated on the *third* day of Passover, the resurrection day of Jesus. "But Christ has indeed been raised from the dead, the *firstfruits* of those who have fallen asleep" (1 Corinthians 15:20).

The first and last days of Passover are solemn assemblies, and no work is done. The *zeroah* (shank bone of a lamb) must be on the Seder plate, representing the perfect, unblemished lamb, whose blood was placed on the center of the lintel and two doorposts protecting Jewish homes in Egypt (Exodus 12:5). The candles are lit, and the Seder begins with the blessing (Kiddush) of the first of four cups of wine:

The Cup of Sanctification - "I will bring you out of Egypt ..."
The Cup of Deliverance - "I will deliver you from bondage ..."
The Cup of Redemption - "I will redeem you ..."
The Cup of Restoration - "I will take you as My people ..."
(Exodus 6:6, 7)

At the Last Supper, it was at the third Cup of Redemption that Jesus said, "This cup is the New Covenant in my blood, which is shed for you" (Matthew 26:27; see Luke 22:20 and 1 Corinthians 11:25). He did not drink the fourth cup. "I tell you, I will not drink from this fruit of the vine from now on until that day when I drink it new with you in my Father's kingdom" (Matthew 26:29).

The Haggadah is the reciting of Israel's deliverance from bondage in Egypt. The children ask four questions, the story of four sons is told, and everyone eats bitter herbs with unleavened bread and *charoset* of four ingredients. The interaction of four and three is symbolic. Three striped, pierced matzot are placed in the *middle* of the table, one above the other. The Seder leader breaks the middle matzo called the *afikomen*.[4] The larger half is wrapped in a linen napkin and hidden. At the conclusion of the Seder, the children search for and find it, and everyone eats a small piece. This represents the Messiah's death and resurrection and recalls Jesus's words: "Unless you change and become like little children, you will never enter the kingdom of heaven" (Matthew 18:3). The Seder concludes with the singing of the Hallel (Praise Psalms; Matthew 26:30).

The civil year begins with the holy day of Yom T'ruah (Day of Blowing), also called Rosh Hashanah (head of the year), on the first of the seventh month, Tishri. It is a joyous but solemn anniversary of the first and the last, the Creation and Day of Judgment. Ten days of penitence are observed during which the Holy One of Israel decides whose names will be written in the Book of Life. Sins represented by breadcrumbs are cast into the waters, a rite called Tashlikh (Micah 7:19). Tradition holds that Adam was created on this day.[17] The shofar is sounded daily, a reminder of the trumpet blasts at Sinai and

---

[4] The Greek word *afikomen* means "I have come." The word *matzah* means "without sin." The idea that the three matzot represent the priest, the Levite, and the Israelite, or Abraham, Isaac, and Jacob, is untenable. When were any of these "without sin" and broken, hidden away, and resurrected?

the ram God provided Abraham. Apples dipped in honey are symbols of the Shekinah and a sweet new beginning.

Yom Kippur (Day of Atonement) is the holiest day of the year, a Sabbath of Sabbaths (Leviticus 23:32). No work is done from sunset to sunset (Leviticus 16:31). All are to humble themselves (Leviticus 23:29). The high priest, cleansed and dressed in holy garments, after atoning for his own sins by sprinkling the blood of a bull on the Mercy Seat *seven* times (Leviticus 16:14), atoned for the sins of the people. "This annual atonement must be made with the blood of the atoning sin offering for the generations to come. It is most holy to the Lord" (Exodus 30:10).

Two goats were chosen by lot, one for the Lord, and one as the scapegoat to carry the nation's sins out into the wilderness. A red sash was tied to its horns. Only blood, a living fluid, can atone for sin.[18] Since there is now no temple altar, there is now no atonement, save for the Blood of the Lamb of God (Hebrews 9:12). Bar-Abbas, whose name means "Son of the Father," was the scapegoat, and Yeshua (salvation), the true Son of the Father, was sacrificed (Matthew 27:20).

Purim is not one of the seven appointed feasts, but its celebration has ritual elements derived from the Assyrian New Year, Sacea. Purim is essentially a spring New Year Carnival. The name is derived from the Assyro-Babylonian *puru* (lot). It commemorates Esther's historical rescue of the Persian Jews from extermination. A Purim king is elected, and gifts are given to the poor. The people wear costumes and masks, and children go from house to house asking for money and treats. Effigies of Haman are paraded through the streets, then hung or burned. Noisemakers are sounded at the mention of his name.[19]

World Memories of the Sabbath Band

Praise Him, you highest heaven, and you waters above the skies.
—Psalm 148:4

The seven levels of Heaven belong to all peoples. How many emergence tales have a royal city surrounded by seven walls? How many have seven hills, seven caverns, seven stories, seven steps, or seven tiers? Sailing the seven seas is a voyage to the heavenly realm. Solomon had seven successive terraces constructed for the temple on which he planted trees. Silbury Hill, England, is formed of seven successive drums of chalk rubble, and Ecbatana, the capital of ancient Media, had seven walls of different colors, one inside the other.[20] In Old Norse tradition, the gates of Valhalla number seven times seventy-seven.[21]

In Buddhist cosmology, seven concentric circles of rock have seven oceans between them, with Mount Sumeru at the center. Chinese buildings in ritual centers were constructed with successive courtyards. On the seventh lunar month, great festivals are celebrated all over China, and the most favored of all amulets is the seven-petaled lotus. The seventh day of the seventh month is the Ghost Festival, when ancestors return and are given offerings. Many countries in the Far East hold similar festivals.[22] The seven levels are recalled as seven tiers in Malasia:

> From the Supreme Being first emanated light towards chaos; this light, diffusing itself, became the vast ocean. From the bosom of the waters thick vapor and foam ascended. The earth and sea were then formed, each of seven tiers. (Walter W. Skeat: *Malay Magic*[23])

In Babylon, the seven cantos of the creation text *Enuma Elish la Nabu Shamamu* (*When on High the Heavens Were Not Named*, circa 1125 BC) were read annually during the New Year festival from the first to the twelfth of Nisanu at the spring equinox. A ritual dethronement and humiliation of the king was followed by his ritual reenthronement.[24] There was general license and disobedience of all commands, and the Zakmuk (feast of lots) was observed. There was a procession through the city of the image of Bel-Marduk to his sacred house, Akitu.[25]

The Irish Tir na n-Og had seven ramparts around its highest building, the great central hall where the high king resided. Each banqueting hall *(bruidhnea)* had seven fireplaces with seven cauldrons containing an ox and salted pig. Each had seven doors with seven roads leading up to it. Outside were seven zones of land and seven of water, giving the appearance of the world mountain.[26] When the kings of Bronze Age Ireland died, their spirits went to a sevenfold spiral castle with a revolving wheel before the door.[27]

Seven represented completeness in Egypt. There were seven gates, seven hawks, a seven-headed serpent, seven scorpions, seven spirits, seven mouths of the Nile, seven Uraei, and seven Arits (divisions of the Otherworld). After death, the deceased addressed seven doors in the House of Osiris.[28] The seventh division contained the Secret Circles of Ament (hidden place).[29]

Plutarch described the ritual "Seeking for Osirus" at the winter solstice when a cow is lead seven times around the temple.[30] The priests invoked the gods by sounding out seven vowels.[31]

Seven is greatly revered in India. It is said in the Rg Veda,

> Seven are the pathways which the wise have fashioned; to one of these may come the troubled mortal. (Book 10, Hymn 5, Ralph TH Griffith, trans. *Hymns of the Rg Veda*

In the Puranas, Brahman creates "seven worlds by seven words."[32] At the pole, Ganga spirals seven times around Brahman's city. In the Vedas, Ganga's source was the highest heaven of Varuna. The Mahábhárata places Ganga in Vishnu's heaven above the polestar Druvha.[33] Agni and Indra are "seven-rayed."[34] There are seven sages, seven rishis, seven islands, seven *lokas* (worlds), and seven centers of pilgrimage. There are seven degrees of Maya (wisdom, supernatural power).[35]

The ancient Greeks had seven Hesperides, seven sirens, and seven satyrs. In *The Iliad*, Homer describes Ouranos (heaven) as a series of enclosures, one above another, the Olympus of many layers or thicknesses, like the curved laminae of a shield.[36] According to Johannes of Lydia (AD sixth century), seven was sacred to Kronos:

> The number seven, as well as his planet, was sacred to Saturn, the god of time; that is Kronos, whom I have identified here with the central polar deity. (*De Mensibus* 25)[37]

The Romans practiced circumambulation of the Ager Romanus (rural area surrounding Rome), marking an invisible barrier against their spiritual enemies. The Salii priests made loud noises and threatening gestures to scare them away.[38] The Circus Maximus was modeled on the celestial prototype. At the sound of a long trumpet, the race began. Four charioteers (*quadrigae*) rounded the *spina* (axial rib dividing the enclosure into two runs) seven times. Turning posts (*metae*) marked each end, and seven eggs counted the seven laps. Each faction had a different color: green, red, blue, and white.[39]

Seven in the Americas

But in the days when the seventh angel is about to sound
his trumpet, the mystery of God will be accomplished,
just as he announced to his servants the prophets.
—Revelation 10:7

Seven signifies completion throughout the Americas.[41] There are seven ancestral towns, seven fireplaces or Council Fires, seven ceremonies, and tribes had seven branches. The Omaha tribe associates the Great Spirit with the number seven:

Ho! Aged One Ecka
At a time when there were gathered together seven persons,
You sat in the seventh place, it is said,
And of the Seven you alone possessed knowledge of all
things.
(Omaha Invocation)[40]

The Hopi trace their origin to *Púpsovi* (seven caves) representing the seven universes, seven stars, and seven worlds.[42] The Zuni bring a yearly offering of corn of seven colors to the priests.[43] The Seven Cities of Cibola are their ancestral pueblos.[44]

Pawnee   Inuit   Moundville   Creek   Mohawk (ES)

The Spanish explorers heard tales of seven cities of gold and massacred many natives to find what was never on this earth. The ancestors of the Nahua (Mayans, Aztecs, etc.) came from Tulan-Zuiva, also called Chicomoztoc (the place of seven caves). The pre-Columbian city of Teotihuacán had a natural cave with seven chambers beneath its Pyramid of the Sun.[45] The Inca tomb of the sun, the *guaca,* has seven steps.[46]

The Labyrinth

Walk about Zion, and go round about her.
—Psalm 48:12a (KJV)

The labyrinth, one of the oldest symbols of paradise, is a diagram of Heaven.[47] Found in temples, churches, and gardens, the labyrinth is a living ritual of humanity's origin and desired destination, with rings spiraling inward toward the center, symbolizing the fall and need for redemption. The ancient root LB (heart) in the holy tongue is *el/ab* (God the Father). We get this root through Latin and Old French (to slip, come loose) hence the English "lapse, collapse, relapse and lapsarian (relating to the fall of man)."[48] Walking the labyrinth was a symbolic return to the Father. The earliest labyrinths are centered in the cross.

Labyrinths were constructed under every palace of antiquity. The classical labyrinth with seven rings is called the "Cretan labyrinth" after the one under the palace of Knossos.[49] All the oldest sky gods were associated with them. The Sumerian Enki (Akkadian Ea) built a labyrinthine temple with a floating garden.[50] As the dead had to pass through a sevenfold maze to reach heaven, megalithic tombs were constructed in that form. The Egyptians believed the dead had to travel a sinuous path to Ament (the hidden land of heaven).[51]

Simple labyrinth (ES)

Labyrinths are found on every continent. Historians mention five great ones: the two Cretan at Knossos and Gortyna, the Greek on the Lemnos, the Etruscan at Clusium, and the Egyptian at Lake Moeris at Hawara, one of the seven wonders of the ancient world.[52] In the Americas, the Inca Island of the Sun had labyrinthine temples built by "the greatest Inca, the ancestor Pachaccuti."[53] The Mayan king renewed his vitality in a labyrinthine palace in Palenque.

From antiquity, pilgrims journeyed to Glastonbury Tor in Somerset in the middle of the Summerland Meadows. Glastonbury is a rare surviving example of a turf or ground labyrinth still in use today. It covers an area of 1,440 acres and was traditionally founded by missionaries led by Joseph of Arimathaea, who built an abbey there. This history is supported by the church fathers Tertullian and Origin. It has been called the English Jerusalem.[54]

Labyrinths in the naves of gothic cathedrals are at the position where Jesus's feet would be if the crucifix were superimposed on the church. At Chartres, it was once possible to read the words of Psalm 51:18, 19 on its stones: "Do good in thy good pleasure unto Zion; build thou the walls of Jerusalem. Then shalt thou be pleased with the sacrifices of righteousness."[55] Walking the labyrinth was a symbolic substitute for a pilgrimage to Jerusalem. The penitent's goal was to escape the material world and find the way home to the Father.[56]

Chapter 8

The Earth of God

Part 1

The Heaven-Earth

For evildoers shall be cut off; but those who wait
on the Lord, they shall inherit the earth.
—Psalm 37:9 (NKJV)

In his commentary on Plato's Timaeus, the philosopher Proclus of Athens made an unusual statement:

> Heaven is in earth, and earth in heaven; but here heaven subsists in an earthly manner, and there earth in a celestial manner. (Proclus on Timaeus trans. by Thomas Taylor)

When God created the earth, He declared it *tov meod* (very good) (Genesis 1:31). The Hebrew word *tov* also means "beautiful, excellent, pleasant, lovely, convenient, fruitful, sound, cheerful." Sounds more like a description of Heaven than earth. We know our early ancestors had some strange beliefs, which we have learned not to dismiss, however incredible they may seem. One of the strangest was locating earth out there in space. Earth was first a cosmic entity. It was the "Otherworld" or the "Netherworld."[1]

In Genesis 1, the creation of earth preceded everything except for the heavens and chaos or void. A celestial earth where God dwells would explain why the earth was created before the other lights of the cosmos. The symbol ⊕ is the standard astronomical symbol for the earth, but it is identical to the sign for the land of God everywhere.

The ancients universally followed Genesis 1:1–2 in the order of creation with earth created before the other celestial lights. They spoke of earth as the foundation that united with heaven, bore all things. The Sumerians spoke of the An-ki (Heaven-earth) encircled by Apsu, the watery deep. The Chinese sages spoke of a united Tien-di 天地 (Heaven-earth). The Egyptian Nu-t (celestial waters, city) was the wife of Geb (the earth god, Creator, and god of the celestial ocean). The Greeks said that Gaea (earth) was married to Ouranos (heaven). Hesiod wrote,

> Verily at the first Chaos came to be, but next wide-bosomed earth, the ever-sure foundation of all the deathless ones who hold the peaks of snowy Olympus. (*Theogony* trans. HG Evelyn-White)

The Irish Tara was originally the earth above, as was the Persian Taera and Hawaiian Taero. The Sanskrit word Tara has fifty-seven meanings, among them "a fixed star, passing over, crossing, carrying across, savior, protector, shining, radiant, pupil of the eye, excellent." The Roman Terra (earth) was united with Coelus (heaven). The Mundus, the circular trench of augurial rites, represented the Otherworld. It is an Etruscan word, which includes this sign: ⊗. It means "the heavens, universe, world, clean, pure, fine and elegant" and is connected with Old High German *mandag*, which means "joyful, happy."

All early tribes had an earth god who lived at the north celestial pole. The Japanese called him "Earthly-Eternally-Standing-Deity."[2] There was a doctrine that the Egyptian Asar, the supreme light of heaven, and morning star was the whole earth and the ocean surrounding the earth.[3] Ptah was called "'Tanen" or "'Ta-tenen," the motionless, resting earth born from the primeval chaos.[4] The Akkadian Ea is called "the Earth Lord" and "Master of the Earth."[5] A title of the god of the Veracruz culture was Tlalxicentica (He Who is in the Earth's Navel). Other symbols were the cross and polestar.[6]

The old traditions were misunderstood after God's earth disappeared, but were not forgotten. Copernicus's sixteenth-century critics feared his revelations of a sun-centered solar system would denigrate earth's geocentric position in *De*

Revolutionibus Orbium Coelestium (1506). Actually, they were both right. Those who held the established view that earth was fixed in the middle of the heavens were right about the former heavens. Copernicus was right about the present temporal arrangement. Had his critics understood the total picture, they would not have forced Galileo to recant his acceptance of the Copernican system, and the former worldview would not continue to be ridiculed as unscientific.

The Stumbling Block

But we preach Christ crucified, unto the Jews a stumbling
block, and unto the Greeks foolishness.
—1 Corinthians 1:23

God's Earth, Four Rivers, and Mount Zion on Mesopotamian cylinder seal.
Artwork courtesy of Ava Raha.

The square, a conspicuous feature of early religious art, is symbolic of the earth and represents divine order.[7] Foursquare symbolism is always applied to our own planet, however incongruous, for no other "earth" is now acknowledged. The phrase "four corners of the earth" often appears in the Bible, "And he shall set up an ensign for the nations, and shall assemble the outcasts of Israel, and gather together the dispersed of Judah from the four corners of the earth" (Isaiah 11:12). "After this I saw four angels standing on the four corners of the earth" (Revelation 7:1a).

Since Heaven was once visible, its square capital, the New Jerusalem, would certainly be remembered. The oldest representations feature a central square, although in point of fact, it is a perfect cube (Revelation 21:16). A cube has twelve lines, eight corners, and six faces. Six directions radiate from the *seventh* central point. The cube at rest is fixed and does not rotate like a sphere. The Holy of Holies was a cube without windows, the only light coming from the menorah in the Holy Place.

A cube-shaped New Jerusalem in the center of the sky is a huge stumbling block to modern comparative mythologists—who with few exceptions ignore it because they can't explain it. It is the same with the terrestrial Jerusalem, a huge stumbling block to humanity's plans for world peace. "And in that day I will make Jerusalem a burdensome stone for all peoples; all who lift it *or* burden themselves with it shall be sorely wounded and all the nations of the earth shall come and gather together against it" (Zechariah 12:3 AMP). The word *block* (AB/El/OK) means "Glory of God the Father" in the holy tongue.

Ezekiel portrays the celestial Jerusalem on a clay block. "Now, son of man, take a block of clay, put it in front of you and draw the city of Jerusalem on it" (Ezekiel 4:1). The Sumerians called it "the Brick of Fate" deposited in the Absu. They described it as "an exact cube" of lapis lazuli.[8] Iku, the Canal Star, was "the star of Babylon." During the fourth day of the New Year creation feast, the high priest stood facing north and recited a prayer three times. "O Canal Star, thou Esagila, likeness of heaven and Earth."[9] The priests then poured over the brick honey and cream.[10] Fifth-century Armenian historian Agathangelos taught that heaven was a solid cube hanging on nothing.[11]

In the center of Mecca is the shrine called the Ka'aba (cube), which dates back to the time of Abraham, who is said to have built it. It was dedicated to an idol of the god Hubal and encircled by 360 images of lesser gods, one for each day of the original year. Before Mohammed destroyed them, Arab worshippers settled disputes by casting lots before the image. Seven arrows were used in this form of divination called Qidh. The most popular form of lots are cubic dice. They have been found in Egyptian tombs.[12] Seven has always been the "lucky" number.

The Swahili-speaking peoples of Tanzania and Kenya describe the Throne of God as a magnificent rectangular carpet. This is the throne of the Last Judgment, where good souls live in joy forever under its shadow.[13] The following partial list of four-cornered objects are prominent in religious symbolism throughout the world:

- the throne-chariot (Merkabah) or car of the gods
- a holy marriage chamber with four-poster bed and canopy
- the public square
- the holy sheepfold or divine cow pen
- a quadrangular castle in the midst of the celestial sea

- a judgment hall, a weaving hall, a banquet hall
- a diamond or cube-shaped star
- a crystal box or casket
- the sacred fields or green pastures
- a square flag, a gem-studded apron
- a fringed "magic" carpet, a woven prayer mat
- a table of the gods (Psalm 23:5, Matthew 22:2)
- a blanket (America), a sheet (Acts 10:11)
- a sail, a great raft or ark
- a magic chessboard

These cubic or square objects are specific to Israel.

The Ark of the Covenant
JewishEncyclopedia.com

The altar of incense
JewishEncyclopedia.com

Temple altar with diamond
JewishEncyclopedia.com

The high priest's breastplate for discerning the will of God
JewishEncyclopedia.com.

Josephus asserted that everything about the tabernacle was made in imitation of the universe. The breastplate in the middle of the ephod of the Kohen haGadol (high priest) represented the earth.[14] The ephod later became an object of idolatry (Judges 8:27, 17:5).

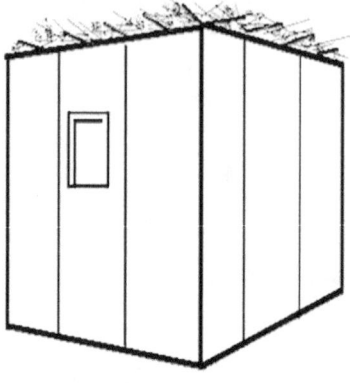

The Sukkah (ES)

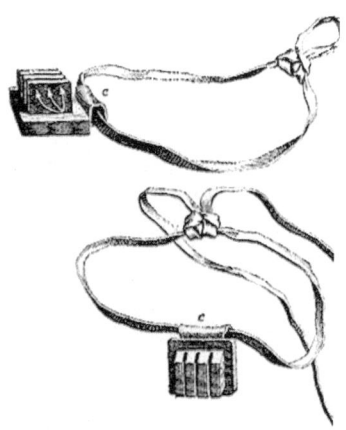

The Tefillin
JewishEncyclopedia.com

The tefillin is a small black box (Shel Rosh) worn on the center of the forehead during morning prayers, with four parallel compartments for verses from the Torah, Kadesh (Exodus 13:1–10), Ve-hayah ki (Exodus 13:11–17), the Shemah (Deuteronomy 6:4–9), and Ve-hayah Im (Deuteronomy 11:13–22). A second black box (Shel Yad) has a leather strap, which is wound seven times around the right arm (Deuteronomy 6:8, 11:18).

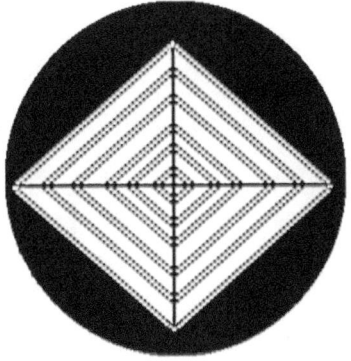

Yarmulke (ES)

The House of God on Earth

And the city is laid out foursquare, shaped like a
cube, and its length is as great as the width.
—Revelation 21:16a (OJB)

The square in Indo-Iranian cosmology has always represented the celestial world.[15] The Persian Khwanirath was the square central space of the seven *keshvares* (regions) of Ahura Mazda.[16] Ahura Mazda warns Yima of fatal winters to come and tells him to build a square Vara:

> Therefore make thee a Vara, long as a riding-ground on
> every side of the square, and thither bring the seeds of sheep
> and oxen, of men, of dogs, of birds, and of red blazing fires
> to be a fold for flocks. Therefore make thee a Vara, long as
> a riding-ground on every side of the *square,* to be an abode
> for men (Zend Avesta, Fargard 2. 25)

Brahman's jeweled city, Devâraka on Mount Meru, is square.[17] Stupas are burial mounds and places of meditation for Hindus and Buddhists. They are built with a square base that represents the seat of Brahman. As the absolute, basic form; no variation is allowed in its construction.[18] Ilâvrita, the city of Shiva, is also "a perfect square" with "no need of the sun or moon to light it."[19]

Ancient Persian square earth (William F. Warren)

Early temples were quadrangular, including those of Mesopotamia, Rome, Greece, and the god-houses on Mesoamerican pyramids. Ancient altars were also square. "Build me an altar of acacia wood five cubits long and five cubits wide; it is to be square" (Exodus 27:1). The Vedic sacrificial fire, the Ahavaniya, which represents the Otherworld, is square and oriented to the cardinal points.[20] The Roman *ara* (altar) was oriented the same way. There was an ara for each cult attached to one square. Only the Aedes Vestae, which represented heaven, was round.[21]

The English architect and historian William R. Lethaby wrote of the prominence of the foursquare earth in the design of ancient temples.

> The perfect temple should stand at the center of the world, a microcosm of the universe fabric, its walls built *foursquare* with the walls of heaven. And thus they stand the world over, be they Egyptian, Buddhist, Mexican, Greek, or Christian, with the greatest uniformity and exactitude. When the world has become circular and spherical, the squareness is retained almost universally as a characteristic of the *celestial earth, the foursquare* enclosure on the *top of the world mountain,* where the polar tree or column stands, and whence issue *the four rivers.* (*Architecture, Mysticism, and Myth*)

The Diamond in the Sky

The Eye of God (El Ojo de Dios) (ES)

He builds his lofty *palace* in the heavens and sets its foundation
on the earth; he calls for the waters of the sea and pours them
out over the face of the land-the Lord is His name.
—Amos 9:6

If the corners of a square face the four directions, they form a diamond. In Mesopotamian art, the diamond had a magical, protective function as a symbol of the earth, the eye, and the womb.[22] The word derives from the Sanskrit *dyu* (luminous being).[23] The Egyptian hieroglyph for *heb* (festival) features a central diamond ⬦. *Heb nefer en pet ta* is "the good festival of heaven and earth." Heb also means "precious stone, hall, garden tent, booth, and tabernacle."[24]

The ancient thread-cross, called "the eye of god," is made with two sticks tied together and strung with colored wool across the frame. It is commonly hung on walls to bless the house. According to author Albert M. Potts, *The World's Eye,* the thread-cross originated in prehistory. It appeared in places as far apart in place and time as Sicily in the third millennium BC, Asia Minor, and South America at the end of the second millennium.[25] In Australia, it is known as the Waningga.[26]

Cherubim guarding the Tree of Life.
Seal Cylinders of Western Asia (William Hayes Ward 1910).

In Chinese lore, four "diamond kings" guard the four corners of paradise.[27] The lord of the diamond is petitioned for blessings and long life in Laos. Heaps of stones lie beside the path, and passersby will add a stone or branch to the pile.[28] In Jewish mystical literature, the Palace of Love is recalled as the place where the Holy One keeps His hidden treasures,[29]

> With the Beginning
> The Concealed One who is not known
> Created the palace.
> This palace is called Elohim.
> (Zohar on the Creation 1. 34a trans. Daniel Matt)

The palace is a symbol of the Unmoved Mover and "the hidden bond which joins man to his Origin and his End."[30] Virgil called it "the lofty palace of the gods,"[31] and Ovid called it "the unkindled palace of the sky."[32] In the Mahabharata, the Gandharvas (celestial singers) live in Gandharva-nagara (castle in the sky).[33] In Welsh tradition, souls at death are conducted to a quadrangular glass castle in the north: Caer Arianrhod, the refuge of Celtic heroes. In its center was Avalon, which shone wondrously at night.[34] In Gwion's poem "The Spoils of Annwn," the castle is called Caer Sidi (Revolving Castle or Castle of the Hill), Caer Rigor (Royal Castle), Caer Pedryvan (Four-cornered Castle), Caer Vediwid (Castle of the Perfect Ones), and Caer Vandwy (Castle on High).[35]

China, the Imperial Capital

Blessed are the meek for they will inherit the earth.
Matthew 5:5

"Heaven is round and earth is square," declare the ancient Chinese philosophers.[36] Shangdi built the first palace, Tiān Zhōng-Gong (Heaven-central-palace), a name for the polestar.[37] *The Book of the Yellow Castle* speaks of "the square foot house, the ancestral land, the heavenly heart, the dwelling place, the face, and the space of the former heaven."[38] Earth is the square frame of a chariot supported by four columns with eight moorings. The sky revolves, but the pure earth in the center does not.[39] The Temple of Heaven is round and the Temple of Earth is square:

> Back in the depths of ancient time;
> Remote, before the Tis (gods) began;
> Four equal sides defined the earth,
> And pillars eight the heaven sustained.
> (*The Texts of Taoism,* volume 2 trans. James Legge)

God's earth is seen in many important characters.
Tiānguó 天国 —kingdom of heaven
Tiān 由 —heaven, sky, field
zhìli 治理 —to govern, rule
bǎoyòu 保佑 or *hēhù*[5] 呵护 —bless, cherish, protect
hàohào 浩浩 —vast, torrential floods
miào 庙 —temple
*hue-*归还 —revert, turn back
shén 神 or *hun* 魂 —immortal soul, spirit.[40]

Hun-t'un (primal chaos) is the virtuous, kindhearted old emperor of the center who died on the seventh day after seven holes were bored into him.[41] His name in Cantonese is wonton.

---

[5] Hehu means "eternity" in Egyptian.

Egypt: The Two Earths

For the pillars of the earth are the Lord's, and
He has set the world upon them.
—1 Samuel 2:8b (NKJV)

In the verse above, the earth is described as having "pillars." This earth does not have pillars, but God's earth does. Certain authors have understood what the ancients were trying to tell us:

> I make this remark in the first place, that we may understand the true sense and importance of those phrases and expressions amongst the Ancients, when they say Paradise was in another world. (Thomas Burnett, *Sacred Theory of the Earth*)

The early Egyptians spoke of two earths in *The Book of the Dead:*

> Hail revolver of Heaven, pilot of the two earths.
> (Chapter 148)
> I have gotten power over the two earths.
> (Chapter 40)

The double glyph ⊗ symbolized God's earth, and Egypt, its mirrored reflection. Taui (earths) is the plural of Ta (earth).[42] The great festival of Samt Taui (Day of the Union of the Two Earths) was a celebration of "the foundation of divine order."[43] Since Ta is also translated "land," this day was later seen as honoring the union of northern and southern Egypt.[44]

The glyph for the Otherworld is ⌒⅃⅂⊗.[45] The glyph *apt* ⊔ (house, palace) represents a quarterly division of the earth above,[46] one of "the four regions of Ra," which are "the limits of the Earth."[47] The letter P □ in Egyptian is a picture of God's square capital. The same character in Chinese is *kow,* a mouth.[48]

Greece: The Pure Upper Earth

When the earth and all its people quake, it is I who hold its pillars firm.
—Psalm 75:3

Plato taught that a far better earth exists above:

Those too who have been preeminent for holiness of
life are released from this earthly prison, and go to their
pure home *which is above,* and dwell in the *purer earth;*
and of these, such as have duly purified themselves with
philosophy live henceforth altogether without the body, in
mansions fairer still, which may not be described, and of
which the time would fail me to tell … Fair is the prize,
and the hope great! (*The Dialogues of Plato,* "Phaedo"
trans. Benjamin Jowett)

Africa: the Dogon

He gathers the waters of the sea together as a heap;
He lays up the deep in storehouses.
—Psalm 33:7 (NKJV)

The Dogon of Mali describe their first ancestor as a great smith who
formed a framework in heaven. It had a circular base and a square top
representing the sky. This edifice was called the Granary of the Master
of Pure Earth. It had seven storehouses, which contained the seeds God
gave to human beings. The granary crashed down to earth on a rainbow,
scattering people, animals, and vegetables. The Dogon village, Ogol, is
oriented north/south. The Dogon house is built in the form of a cross with
the door opening to the north.[50] They venerate an invisible star, "the most
important star," called Po tolo, which means "deep beginning," and mourn
the death of the First Ancestor in the public square every sixty years.[51]

America: The Sky-Earth

Early kivas, the underground temples of the Hopi and other Pueblo tribes, were constructed in the heavenly pattern. Oriented to the four directions, they had a conical roof supported by four great masonry pillars and a square entrance facing the North Star. Sunk in the navel of the village, there is a fire pit directly below the roof opening. Beside it is a small hole, the Sipapuni, meaning "navel" and "path from." Here the New Fire ritual Wuwuchim is celebrated. It is said: "man is rooted in this world until he is planted on another at a future time."[52]

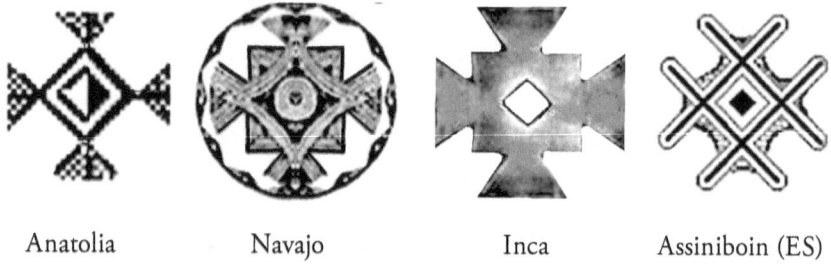

Anatolia   Navajo   Inca   Assiniboin (ES)

The Maya viewed the cosmos as a giant square earth upon which human beings walked. In the center was the holy ceiba or yaxche tree whose branches reached to the heavens.[53] In the Popul Vuh, the Mayan Cahuleu (Sky-earth) is described as *cah tzuc* (four-sided) and *cah xucut* (four-cornered). The square in the circle symbolizes the Creator Hunab Ku (Sole God). His home had many names: "the World Dwelling," "the Great Lodge," "the House Made of Dawn," "Star House," "Sweat Bath House," "Talk-Place House," and "Celebrated Sea-House."[54]

The South Pacific

He then brought me to the outer court and led me around to
its four corners, and I saw in each corner another court.
— Ezekiel 46:21

To native Australians, a flat earth was present in the dreamtime concave sky. It was a rich country with plenty of water. All the heroes and ancestors lived there. The stars are seen as their campfires. Earth is the foundation of all life and the place from which their ancestors arose. The elements of earth are associated with the four sacred colors, red, black, yellow, and white. The Kooris, the indigenous people of New South Wales and Victoria, who speak the Awabakal tongue, believe the sky was raised on props at the corners of the earth.[55]

The traditional kite, like the thread-cross, dates to prehistoric times. As a picture of the sky-earth, it has many mythological associations. Kites were used to send messages to the gods. The Polynesian cultural hero Maui stopped the sun's movement. When his kite was in the sky, "The weather was fine."[56] Kite flying is the central activity in spring New Year feasts in many places. In India and Pakistan, kite celebrations are called Basant or Vasanta, Sanskrit for "spring."

1635 woodcut, author: John Bate

Mesopotamia: The Holy Sheepfold

Son of man, this is what the Sovereign Lord says to the land of Israel:
"The end! The end has come upon the four corners of the land!"
—Ezekiel 7:2

On cylinder seals, the Sumerian Enki (Akkadian Ea) sits on a square throne in his square sea-house. As the founder-king, four rivers flow from him. Enki means "Lord of Earth," but he was also Sar apsi (king of the Absu).

Enki cared for the poor, weak, widowed, and wronged. As the benefactor of humanity, he was savior, healer, counselor, and provider. He organized the four-cornered earth and established law and order.[57] One of his titles was Tirannu (rainbow),[58] Enki's city was Uruk (biblical Erech Genesis 10:10.) Uruk means "pure house, holy sheepfold," At the Flood, Enki's seat became a deserted. sheepfold.

Ea sits on his square throne. Four Rivers flow from him. (Morris Jastrow 1911)

Roma Quadrata

The earth is the Lord's and everything in it, the world, and all who live in it; for he founded it upon the seas and established it upon the waters.
—Psalm 24:1, 2

The city of God was the prototype of early Rome. Mircea Eliade recognized the pattern as a square within a circle. "The twin tradition of the circle and the square was so widespread as to suggest it."[59] The Romans divided their territory into four *argea* around the Forum, regions that constituted the Roma Quadrata (Square Rome).[60] North of the Forum was the Comitium, the political center and place of assembly, an inaugurated plot oriented to the cardinal points at the foot of the Capitoline. In the center of the north side was the *curia,* a quadrangular building where the senate held council.[61]

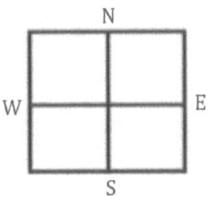
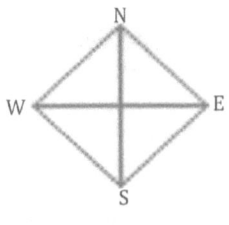

Roman   Umbrian.[62]

The Mandala

A square plot of five hundred by five hundred cubits shall be for
the sanctuary, with fifty cubits for an open space around it.
—Ezekiel 45:2b

In Hindu philosophy, "Completeness has four angles and is supported
on four feet."[63] The four-cornered earth is mentioned in the Vedic hymns,

Thy spirit, that went far away to the four-cornered earth …
to the four quarters of the world. (*Rg Veda* book 10 Hymn
58, Ralph TH Griffith, trans. *Hymns of the Rg Veda*

Jain, Hindu, and Buddhist cities are built as mandalas (Sanskrit for circle).
A series of concentric circles surrounds a square central shrine, representing
Mount Meru.[64] Mandalay was one of the last great cities laid out in the
canonical design. Mandalas are incorporated into shrines and gardens. Every
temple seen from above is a mandala. Tibetan monks painstakingly recreate
them every day. The contemplation of a mandala is an aid in meditation.[65]
All these intricate works of art are then destroyed and reverently returned to
the Creator of the original. The yogi attempts to place himself mentally into
a mandala in order to attain union with God.

Kalachakra,   Native American (ES)

Among Native Americans, the mandala is the mandela, a medicine wheel used as a wall hanging and symbol of blessing. The mandela with eagle feathers is a symbol of the Great Spirit and the fountain of living waters.[66] Cities throughout Mexico and Central America, like Teotihuacan and the Aztec Tenochtitlan, were laid out in this heavenly pattern.

The Public Square

> Justice is turned back, and righteousness stands at a distance; for truth stumbles in the public square and uprightness cannot enter.
> —Isaiah 59:14

Every tribal settlement on earth once had a central public square or green commons. These squares, modeled on the heavenly prototype, were used for worship, justice, and assembly. The Russian Arbat is the ancient navel of Moscow. It is called Arbatskaya square (Арбатская площадь). In the holy tongue, Arba means God the Father. In Hebrew, *arba* means "four." Without the *aleph,* the word *raba* means, "to have four sides, to be square." In Greece, the public square is the *agora*. Author John O'Neill writes:

> The Agora was the celestial place of assembly of the gods, whence the Word of God proceeded, before it became the earthly meeting-place of men where their debates took place. The archaic Agora, like the Roman Forum, was the very center and heart of the city. It was rectangular, in the form

of a plinthos or brick. (John O'Neill, *The Night of the Gods,* Volume I (1893)

The first structure erected in a tribal settlement speaking the same tongue was the temple. It was the buying and selling of sacrificial offerings that turned its central crossroads into the marketplace. Fairs and circuses, as temporary returns to the golden age, are laid out in the heavenly pattern, as are royal courts. It was formerly the practice in the common law of northern Europe to create a square enclosure laid out as a grid as a place of jurisdiction, a practice surviving in Germany until the eighteenth century in the illegal summary courts of Westphalia.[67]

A New Earth

Now I saw a new heaven and a new earth, for the first heaven and
the first earth had passed away. *Also there was no more sea.*
—Revelation 21:1

The Inuit of Greenland say that after the blessed time of no night and no death, earth fell from heaven, and the Flood came upon us.[68] The Word of God promises us a new heaven and a new earth for the righteous. "The righteous will inherit the land, and live in it forever" (Psalm 37:29). Who is righteous? "There is none righteous, no, not one" (Romans 3:10), (Psalm 14:2, 3). If only the righteous can inherit the earth, what hope is there for the children of Adam? This is our hope: "For my Father's will is that everyone who looks to the Son and believes in Him shall have eternal life, and I will raise him up at the last day" (John 6:40).

Creation was completed with the first Adam. The last Adam, Jesus Christ, completed our redemption. "For since death came through a man, the resurrection of the dead comes also through a man. For as in Adam all die, so in Christ all will be made alive" (1 Corinthians 15:21, 22). What was lost through the first Adam's disobedience was regained by the last Adam's obedience. "For just as by the one man's disobedience the many were made sinners, so by the one man's obedience the many will be made righteous" (Romans 5:19).

Part 2

The Emerald City

> And He who sat there was like a *jasper* and a sardius stone in appearance;
> and there was a rainbow around the throne, in appearance like an *emerald*.
> —Revelation 4:3 (NKJV)

On the Isle of Patmos, John saw God's throne encircled by a rainbow glowing like an emerald. What was the source of this beautiful light? Do other cultures recall it? In rabbinic tradition, Tohu was a green band that encompassed the whole world.[1] In the Midrash Konen, this band "mediated between the edges of Heaven and the edges of the earth."[2] Tohu is the singular of Tehom (deep) of Genesis 1:2a, "The earth was without form, and void; and darkness was on the face of the deep."

Throughout the world, God and His city were represented as green. Emerald walls surrounded Elyseum,[3] and woven green laurel branches formed the victory crown. Wherever they settled, the people looked for green and turquoise minerals to represent their Creator. Jasper, a highly prized green stone (Greek *iaspis*, Hebrew *yashpheh*, Persian *yashp*), begins with His name. Emeralds were believed to promote health, wealth, cure leprosy, eye disease, and snakebite. Their possessors were said to be under God's protection and endowed with foreknowledge.[4] Green, the color of life, is the middle color of the rainbow. In the ancient mystical science of alchemy, the sign for green is a cross within a circle.[5]

This was the case throughout Egypt in a long white robe with his face and hands painted green. He is sitting on a square throne above a four-cornered lake with a water lily (lotus) growing out of it. Before him stand four Kherefu (cherubim) as guardians. In hymns, he is described as beautiful,

> Thy body is of gold, thy head is of azure,
> and emerald light encircles thee.
> —*Book of the Coming Forth by Day*[6]

An iridescent green scarab beetle and a mouth were the hieroglyphics of Khepera, a form of Ra worshipped in Heliopolis,

The name derives from *kheper* (to make, to produce) who said, "I was the Creator of everything which came into being … which came forth from out of my mouth."[7]

Sebek-Ra-Temu, the god of Ombos and Nubit, was called "the beautiful green Aten, which shineth ever, the Creator of whatsoever is and of whatsoever shall be."[8] A hymn to Ra declares,

> They give a shout of joy at thy coming forth. [Thou] in the *Khu-t*[6] of Heaven, thou sheddest upon the taui (earths) *emerald* light. (*The Egyptian Book of the Dead*, chapter 15)

A hymn to Ptah-Tanen speaks of the Great Green Sea,

> Moulder of gods and of men, and everything, which is produced, maker of all lands, and countries and the Great Green Sea … [9]

Green scarab amulets set in gold and silver were placed on the breast of the deceased and worn as jewelry.[10] The Egyptian royal scepter *uat* (verdure, greenness) was constructed of matrix emerald.[11]

The color green is associated with many tribal deities and heroes throughout the world. Sir Gawain wears a magically protective green girdle and fights with a green knight.[12] Childe Rowlande climbs a green hill surrounded by rings from the bottom to the top. He goes around it three times saying, "Open door." When he enters, the door closes. Within it is a treasure house with a diamond keystone.[13]

Herodotus wrote of the temple of Melqart (king of the city) at Tyre, which held a colossal emerald described as a fragment of a fallen star.[14] The Sumerian Enki (Akkadian Ea) was pictured with a green face, and his heaven was adorned with jasper.[15] The mystical yearning for the city of God was transposed to one's own land. As the Irish longed for Erin, the Emerald Isle, the ancient Persians yearned for Aîrân with its emerald rocks and emerald cities in a green island at the pole (Kutb).[16] The worldwide associations between the city of God and the color green (or blue-green) are inexhaustible.

---

[6] *Khu-t* (palace) is related to *khu* (to rule, spirit, soul, and Aakhu (the Great Light). Also related is the Hebrew Mal*khu*t (kingdom) and *samkhut* (authority), the Persian Kutb (pole). *Ku* is "holy" in Mayan and "ancient" in Chinese.

The Jade Emperor

(He) showed me the Holy City, Jerusalem, coming down out of
heaven from god. It shown with the glory of god, and its brilliance
was that of a very precious jewel like a jasper clear as crystal.
—Revelation 21:10, 11

In China, green imperial jade symbolized Shangdi, "the August Supreme
Emperor of Jade, who created men from clay."[17] The character for jade 玉 (yù), a
cross between heaven and earth, is identical to the one for king 王 (wáng), except
for one stroke. Yao, an early epithet of Shangdi, had a green face, and his glory
filled the empire. Holy objects, including a green tablet, the *kuei*, were placed in
imperial coffins in an arrangement representing the brilliant cube 方冪 *fang ming*,
an emblem of sacrifice.[18] The word Yao has many definitions, among them "to
shine, green jasper, remote, mysterious, retired, glorious, key, medicine, an ancient
emperor, and a sacred mountain."[19]

The emperor's personal symbol was the jade *bi*, a round disk with a central
hole identified with the polestar. The Taoists called it "the Pivot of Jade."[20] Some
have been found dating to the Neolithic period. The emperor used the bi to
commune with heaven.[21] Jade chimes were used in religious ceremonies, and like
the Egyptians, jade amulets called *han-yü* (mouth-jade) were placed in the mouths
of the deceased.[22]

The old character for earth, *tǔ* 土, is the sign of God on earth. In both
Chinese and Japanese belief, the planet Saturn 土星 is the earth star (see Amos
5:26 AMP). The Kojiki speaks of the compassionate great ancestor Oho-Kuni-
nushi-no Kami (Master Deity of Great Land) who rules over the hidden earth
on Mount Mimoro (volume 1, 106). He disappeared in "the fence of the *green*
branches of Yamato," a much longed-for place,

> Eight clouds arise.
> The eightfold fence of Idzumo makes an eightfold fence
> for the spouses to retire within.
> Oh! that eightfold fence.
> (Kojiki Volume 1)

This may be another way of saying "behind the lattice" (Song of Songs 2:9).
Another name for paradise is Todaru-Ame-no-Misu (plentiful heavenly dwelling).
Misu can also mean "lattice."[23]

In Vietnam, the Jade Emperor is known as Ngoc Hoang. Earth is a square plateau, and His earthly temple is in Ho Chi Minh City (Saigon).[24] The supreme god of the Bonpos of Tibet, Nyatri Tsenpo, lives in a gold castle with a turquoise roof in the seventh level of heaven. Manosawar, the Bathing Lake of Tibetan lore, resembles a turquoise mandala. Its waters are covered with rising blue smoke. There are jeweled swamps, golden sands, and coral plants. At its center is a weeping plant shaped like a diamond with a sweet-scented, eight-petaled turquoise lotus. Covered with dew and sparkling with lights, it is said to cure all ills.[25]

The colors green and turquoise were auspicious throughout the Americas. The Navajo say when the heavens were close, there was a great light from the square turquoise House of the Sun, which stood on mighty water.[26] The Caranques of Ecuador made pilgrimages to a large emerald that represented Umma, their deity of medicine.[27] The Maya enclosed their sovereign in blue-green quetzal feathers (Yayax).[28] Quetzalcoatl's beautiful city of jade, Tula, had "green cross-beams."[29] The Veracruz Huehueteotl (Old Old God) was "Lord of the Turquoise." He held a gold buckler with five jadeite stones in the form of a cross on a gold plate. Ordinary people were forbidden to wear or own turquoise, as it was the property of the gods.[30]

The Emerald City was the destination of the four main characters of *The Wonderful World of Oz*. The author, Frank Baum, was an ardent student of theosophy. He drew the basic elements of the story—the Emerald City on the other side of the rainbow, the spiral yellow brick road, and the witches of the four directions—from ancient lore. To the good and beautiful witch of the north, he gave a wand with a star. Aside from political satire, his exposure of the kindly old wizard as a blowhard in a hot-air balloon reflected his opinion of religion. The word *oz* in Hebrew means power, "Ascribe ye *oz* unto Elohim; His majesty is over Yisroel, and His *oz* is in the skies" (Psalm 68:34 OJB).

The Lake of Life and the Well of Salvation
Egyptian Lake of Life (EAW Budge)

With joy you will draw water
from the wells of salvation.
—Isaiah 12:3

The ancients tell of a crystalline lake in the center of God's earth as the prototype of all sacred lakes and the sites of popular pilgrimages. The pool of Siloam where the blind man received his sight (John 9:7) and Beersheba (well of seven oaths) are biblical examples. As they were seen as gates to His kingdom, votive offerings were cast into them. Tossing a coin into a wishing well is a vestige of this ancient practice. Author William Lethaby comments on the source of these worldwide traditions:

> And men there beyond say that all the sweet waters of the
> world above and beneath take their beginning from the well
> of Paradise, and out of that well all waters come and go.
> (WR Lethaby, *Architecture, Mysticism, and Myth*)

God's name is the root of the English words: *lake, pool,* and *well* (HW/EL). Some cultures, such as the Celts, practiced circumambulation of wells.[31]

The Chinese garden of Kunlun, with its flowery platform, had a Lake of Gems on its summit.[32] The Chinese glyph for *jing* (well) is 井. Eight families farmed the surrounding fields. The center plot with the well, called "God's Acre," was worked in common for the emperor.[33] Here is the ancient pound sign, which belongs to the king and came to be used for a jail (e.g., dog pound). In many board games, the central square of the quadrangular board is for impounded pieces.[34]

Lake Titicaca is sacred to the Peruvians as their place of emergence. The Collao people call it Mamacota (Mother-water). They placed a Copacahuana—an idol of bluish-green stone shaped like a fish with a human head—in a commanding position on the shores of the lake. The worship was so deeply rooted that the Spanish friars could only suppress it by replacing it with an image of the Virgin Mary.[35]

Mimir (also called Hvergelmir) is the bubbling Norse celestial well of mead, where Odin sacrificed his eye for wisdom.[36] The Persian fountainhead Ardvî-Sûra in the garden of Ahura Mazda is the mother of all terrestrial waters and source of all good things.[37] In Arab tradition, the Well of Souls, *bir al-Aruah* is said to lie under the Dome of the Rock, where since the establishment of Islam, it is said the Twelfth Imam awaits his time of return. These beliefs existed thousands of years before Mohammad.[38]

After the Flood, the ancients believed that the celestial earth with its central Lake was now the underworld. To this fiery lake come the souls of the dead. According to the Greeks, this is the celestial world's dark reverse ruled by Hades (called the subterranean Zeus). The four rivers flowing in and out of it are now the River of Pain (Acheron), the River of Burning Fire (Pyriphlegethon), the River of Hate (Styx), and the River of Wailing (Cocytos). The outermost river, Okeanos, flows in a circle around the whole dark invisible world.[39] According to the Egyptians, the life of the land died away when the god departed from his city and resided in Edin, the Underworld.[40]

The Green Fields and Pastures of the Lord

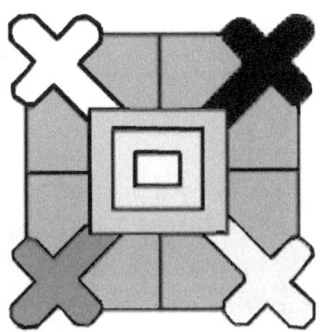

Zapotec earth with four green pastures and central lake, Mexico (ES)

He makes me lie down in green pastures, he leads me
beside the still waters. He restores my soul; He leads me
in the paths of righteousness for His name's sake.
—Psalm 23:2–3 (NKJV)

Titles for heaven often contain the words *fields* and *plains*. The Greek Elysian Fields and the Irish Mag Fail (Plain of the Field of Fal) are examples. These are divided into four green pastures. In the Pyramid Texts, Horus (Heru, "the face") is called "Lord of the four Greens."[41] Horus in Nekhen had the title of "Lord of the Fields of Emerald."[42] The sacred lake is always in the exact center:

> Pure, pure is the Lake of Aaru. Pure is Ra in the Lake of
> Aaru. He brings the crystal of the Great Eye in the midst of
> the field. (Pyramid Texts of Pepi I trans. EAW Budge)

The Yakuts of Siberia describe a Lake of Milk in the midst of a green, grassy plain over which no wind blows.[43]

The heavenly fields are the prototype of our own fields, making the acts of plowing, sowing, and harvesting holy. In Israel, the four corners of a field were sacred and left for widows and the poor (Leviticus 23:22). The Roman feast Ambarvalia was the name given to the ritual purification and protection of the *campus* (field). In this rite, it was believed an invisible spiritual barrier against human and demonic enemies could be established by parading the *hostia ambarvalis* (victim) three times round the fields, before the sickle was put to the corn. Marching behind the victim was the *chorus et socii,* a crowd of merrymakers—reapers and farm servants— dancing and singing before the victim was slain.[44] This was done for the *urbs* as well as the *ager.*

The people of Japan use the Chinese character 田 (field, heaven) for Ta, their "Enclosed Rice-field" symbolized by green jade. The Nihongi calls this Ashi-hara no naka tsu Kuni (the Central Land of the Reed-plain). Ashi is the axis and Hara is the plain.[45] Shinto purification rites involve walking through a circle of rope, a large sacred ring made of reeds called a *chi-no-wa.*

The Babylonian deity Bel-Marduk, after whom Mordecai was named, covered the central portion of the circle of primeval waters with a "great reed frame filled with earth" to make a place for the gods and men. His anger was the cause of the Flood.[46] The Norse Idavollr (whirl-field) was the green plain in the center of Asgard.[47] At Ragnarök it became a vast battlefield where the gods were defeated and disappeared. The earth lost its shape, sank beneath the sea, and could no longer be seen.[48]

The Zulu creation account tells of the Ancient One, Unkulunkulu. No one knew where he came from. It was said that he broke off the people and the animals from some reeds. The word used for reeds, *uthlanga,* also means "source."[49] Rushes and reeds are common in ritual, because they grow near rivers. The baby Moses was placed in a little ark of rushes. Green jade represented the Toltec paradise of Tolan, which was called "the Place of Reeds." The date name of its first ruler *Ce Acatl* (the morning star), who was older than the sun, was "One Reed."[50]

Old songs and rhymes contain hidden truths. The popular English folk song "Green Grow the Rushes O" always returns to the God One:

> I'll sing you One, O
> Green grow the rushes, O
> What is your One, O?

Green grow the rushes, O
One is One and all alone
And evermore shall be so.
(Anonymous)

The Egyptian Word Sekhet

He told them another parable: "The kingdom of heaven is like a mustard
seed, which a man took and planted in his field. Though it is the
smallest of all seeds, yet when it grows, it is the largest of garden plants
and becomes a tree, so that the birds come and perch in its branches.
—Matthew 13:31–32

All Egyptians hoped to ascend to the Sacred Fields where the souls of the
blessed dead served Asar (Osirus). Arrayed in white linen garments and white
sandals, they expected to eat luxuriant crops unknown on earth and longed
to sit by the lake where the Tree of Life grew:[51]

> He travelleth to the great lake in Sekhet-hetep by which the
> great gods alight, and these great ones of the imperishable
> stars give unto Pepi the Tree of Life whereon they themselves
> do live, that he also may live …
> (Pyramid Texts[52])

For the righteous, it was a Lake of Life, but for the enemies of Asar, it was a
Lake of Fire[53] (see Revelation 20:15).

Sekh-t (field, garden, meadow) begins many names for heaven:

- Sekh-t neheh (Field of Eternity)
- Sekh-t ankh (Field of Life)
- Sekh-t-aaru (Field of Reeds)
- Sekh-t uatch-t (Emerald Fields)
- Sekh-t mefkat (Turquoise Field)
- Sekh-t-hetep (Field of Peace)
- Sekh-t Ra (Ra's dwelling)[54]

For the Egyptians, ritual was all about getting back to the Sacred Fields. Here is the link to the biblical Feast of Sukkot. A related Egyptian word—*skhet*—means, "to erect a shelter made of leaves and branches, to weave, to build a booth."[55]

Sukkot, the Eternal Feast

And every one that is left of all the nations which came against Jerusalem shall even go up from year to year to worship the King, the Lord of hosts, and to keep the feast of tabernacles.
—Zechariah 14:16

The Egyptian word Sekh-t is very similar to the Hebrew word Sukkot. The singular *sukkah* (pronounced *soo-kah)* originally meant "woven."[56] Until the destruction of the temple, every able son of Abraham made a pilgrimage to Jerusalem to celebrate Israel's final feast in the middle of the seventh month. Sukkot may be the oldest sacred feast in the world, going back even earlier than the time of Abraham. This was a field and harvest festival and the most prominent of all, for it is declared to be "the Feast of the Lord" (Leviticus 23:39; Judges 21:19), "the Feast of Ingathering" (Exodus 23:16, 34:22), or just "The Feast" (1 Kings 8:2; 2 Chronicles 7:8). Days one and eight, the *azeret* (crown) are "Sabbath-rests" (Leviticus 23:39).[57]

We learn from the Mishna the drama and symbolism in the way Sukkot was celebrated at the time of Christ with traditions featuring light, water, and seven circuits. The book of Jubilees records Abraham celebrating it in a similar way:

And Abraham took branches of palm trees and the fruit of good trees, and every day going around the altar with the branches seven times in the morning, he praised and gave thanks to his god for all things in joy. (Jubilees 16:3)

The ceremony of the water drawing (or water-libation) was performed each morning. The Levitical priest descended to the pool of Siloam accompanied by flutes. The people then entered the temple through the Water Gate while the ram's horn trumpets were sounded. The priest placed two silver basins on each side of the altar. Into the one on the eastern side, he poured the wine of the drink offering. Into the western one, he poured water from the pool of Siloam. This represented the pouring out of the Holy Spirit (Joel 2:28, Isaiah 12:3).[58]

At the end of the first day, four enormous lamps fifty cubits high with golden bowls were set up in the quarters of the outer court. Four ladders leaned against them, and the priests and Levites, using their worn-out liturgical clothing for wicks, filled them with pure olive oil. All night long, joyful songs of praise and thanksgiving were sung to the Lord, and the people took part in a torch dance. The Levitical musicians and choir stood on fifteen steps corresponding to the fifteen psalms of ascent (Hallel). The temple in Jerusalem illumined the night like a diamond. It was a holiday of great *simcha* (joy).[59]

The seventh and last day was "the Day of the Lord." Two ceremonies were conducted: the water-libation and the Hosha'na Rabbah (the Great Hosanna), the torch-lit seven circuits of the altar. The people cried out "Hosheanah!" (Deliver us now!) seven times while marching around the altar, carrying palm branches. Willow branches were set up at the sides the altar.[60] An old tradition holds that Jesus was born on this feast and circumcised on the eighth day. It was on the last day of the feast that He announced who He is and His mission,

> The Spirit of the Lord is on me, because he has anointed me to preach good news to the poor. He has sent me to proclaim freedom for the prisoners and recovery of sight for the blind, to release the oppressed, to proclaim the year of the Lord's favor. And he rolled up the scroll, gave it back to the attendant, and sat down. The eyes of all in the synagogue were fixed on him. Then he began to say to them, "Today this Scripture has been fulfilled in your hearing." (Luke 4:18–21)

On the last day, that great *day* of the feast, Jesus stood and cried out, cried out, saying, "If anyone thirsts, let him come to Me and drink. He who believes in Me, as the Scripture has said, out of his heart will flow rivers of living water." But this He spoke concerning the Spirit, whom those believing in Him would receive; for the Holy Spirit was not yet *given,* because Jesus was not yet glorified. Therefore many from the crowd, when they heard this saying, said, "Truly this is the Prophet." Others said, "This is the

Christ."(John 7:37–41 NKJV) Jesus is the fulfillment of the Feast, "And the Word became flesh and dwelt (or tabernacled) among us, and we beheld his glory, glory as of the only begotten son of the Father, full of grace and truth" (John 1:14 NKJV).[61]

There are several Hebrew names in the Bible for the day when the earth will be purified by fire (2 Peter 3:7,10):

- Yom HaAcharon—last day
- Yom Ha Din—day of judgment
- Yom Pekuddah—day of visitation, reckoning
- Ketz HaYamim—end of days
- Yom Ra'ah—day of evil
- Yom Hashem—day of the Lord
- Yom Af—day of wrath

Psalm 27 in the Orthodox Jewish Bible enlightens and comforts those who hope in the Lord,

> For in the *Yom Ra'ah* (day of evil) He shall keep me safe in *His Sukkah;* in the shelter of His *Ohel* (tent) shall He conceal me; He shall set me up upon a *Tzur* (rock). And now shall mine head be lifted up above mine enemies round about me; therefore will I offer in His *Ohel* sacrifices with shouts of joy; I will sing, yea, I will make music unto *Hashem* (the Lord). (Psalm 27:5, 6 OJB)

Chapter 9

The Garden of God

The Search for Eden

The Lord God took the man and put him in the
Garden of Eden to work it and take care of it.
—Genesis 2:15

Where is the Garden of Eden? Does it still exist somewhere? Some believe the story to be an allegory, but the ancients did not share this view. Others assumed that God planted His garden on earth as hallowed ground and all traces of it were obliterated at the Flood. Some believe it was in Iraq between the rivers Tigris and Euphrates. Others have thought it lies at the bottom of the sea or is frozen under a glacier. Fortunately, we can know where Eden is because the Bible tells us exactly where it is.

Eden's Four Rivers at the North Pole, 1623 Gerardus Mercator Atlas ⚠.

After a lifetime of studying ancient texts, author and scholar William F. Warren concluded that Eden was the center of an antediluvian continent at the North Pole.[1] Indeed, world tradition does locate the cradle of the human race at the pole, and the remains of many tropical plants and animals have been found quick-frozen inside the Arctic Circle. In the Midrash, however, there are two gardens, one in heaven known as the "higher garden," the habitation of righteous souls, and the "lower garden" on earth, the upper garden being the prototype of the lower.[2] According to the *Jewish Encyclopedia*,

> The Talmudists and Cabalists agree that there are two gardens of Eden: one, the terrestrial, of abundant fertility and luxuriant vegetation; the other, celestial, the habitation of righteous, immortal souls. These two are known as the "higher and lower" Gan Eden.[3]

Some Rabbis taught that Eden was preserved during the Flood on a mountain reaching to the moon,[4] a tradition that exists among many peoples. Albertus Magnus (1193/1206–1280) wrote:

> I have found it in some most ancient books that Thomas, the Apostle, was the author of the opinion ... that Paradise was so high as to reach to the lunar circle. (*Summa Theologiæ*)

Ezekiel locates Eden on the Mountain of God and speaks of Lucifer:

> You were in Eden, the garden of God: every precious stone adorned you: ruby, topaz and emerald, chrysolite, onyx and jasper, sapphire, turquoise and beryl. Your settings and mountings were made of gold; on the day you were created they were prepared. You were anointed as a guardian cherub for so I ordained you. You were on the *holy mountain of God* you walked among the fiery stones. You were blameless in your ways from the day you were created till wickedness was found in you.
> (Ezekiel 28:13–15)

Landmarks of Eden

And a river was going out from Eden to water the garden; and
from thence it was divided, and became four heads.
—Genesis 2:10 (KJV)

According to Genesis, Eden's four rivers are the Gihon, Pishon, Hidekkel,
and Phrat. Could two of these be Iraq's Tigris and Euphrates? The early
Mesopotamians believed these rivers had a common source in the far north,[5]
but nothing in the Sumerian or Babylonian texts indicated that they believed
their country was the cradle of the human race. Not one of the early Christian
fathers, either Latin or Greek, placed Eden anywhere in Mesopotamia.[6]

Where then are the four head rivers of Eden? The Bible places them
in the City of God, a divine picture fixed in human memory. "There is a
river, the streams whereof shall make glad the city of God, the holy place of
the tabernacles of the most high. God is in the midst of her; she shall not
be moved" (Psalm 46:4, 5 KJV). This is the Navajo To Alnaozli (Crossing
of the Waters) or To Bil Dahisk'id (Place Where the Waters Crossed).[7] The
Four Niles of the Neter Ta (God Land) are seen in the glyphs for the crowns
of Egypt.

Crowns of Upper and Lower Egypt (EAW Budge)

When arriving in a new territory, it is common practice to name local
landmarks after those in one's homeland. A famous spring in a cave in
Jerusalem was named after one of the rivers of Eden, the Gihon. Temples
and cloisters all had their paradise gardens. There were certain features of
Eden, however, that could not be part of the local landscape, the Ever-turning
Sword, the Tree of Life, and the Cherubim, although some shrines had
representations of them. The Sphinx that guards the Valley of the Kings in
Egypt is a type of cherub.

Eden inspired the crosswalks of our traditional cloister gardens,

Monestir de Santa Maria de Ripoll

The first and last books of the Bible tell us the location of Eden and the two trees: "In the middle of the garden were the tree of life and the tree of knowledge of good and evil" (Genesis 2:9b). Revelation 2:7 locates the tree of life in Heaven. "To him who overcomes, I will give the right to eat from the tree of life, which is in the Paradise of God" (Revelation 2:7; see Revelation 22:2, 22:14). There is no contradiction here. The Hebrew term for Heaven is Gan Eden (the Garden of Eden). The Arabs call it Jannat Adn (fixed residence) the eternal abode of the faithful.[8]

In the floor of the Dome of the Rock, a square, green, jasper slab once lay. The Arabs called it the Stone from Eden. The north gate of the mosque nearest this stone is called the Bab ej-Jinah (Gate of the Garden of Eden). Mohammed was reputed to have driven nineteen nails into it. The Arab judge Mujir ed-Din, who lived in Jerusalem in the fifteenth century, said that Solomon's grave lies beneath it. During World War I, Jamal Pasha, commander of the Turkish forces in Palestine, removed the stone, and its location is presently unknown.[9]

The Mystery of Kedem

And the Lord God had planted a paradise of pleasure *from the beginning:* wherein he placed man whom he had formed.
—Genesis 2:8 (DRA)

In 1885, William Warren pointed out that only one Hebrew word in the entire Bible connects the Garden of Eden with the east—*kedem:*

How is it that with such perfect unanimity on the part of contemporary nations in respect to the north-polar position of the cradle of mankind, the traditions of the Hebrews alone would have placed it in the East?
(*Paradise Found*)

Others have also asked this question:

The obscure *meqedem* after "And Yahweh Elohim planted a garden in Eden means 'in primeval times', as so often in the Hebrew Bible, and not 'from the east.'
(William F. Albright, *Yahweh and the Gods of Canaan*)

The word kedem only became associated with the east after the Flood and a long period of darkness. It was only natural that the first sunrise would assume a highly religious significance on an independently rotating earth. The definitions of kedem are "what is in front, before" "a primitive or ancient state," or "primeval." The contracted preposition *me* is "away from." One good answer is, "What does *me-kedem* mean? Away from the Ancient One" (Zohar 1:76b).

In 1691, Thomas Burnet wrote in his *Sacred Theory of the Earth*:

Some have thought the word *me-kedem* was to be rendered "in the East, or Eastward," as we read it, and therefore determined the site of Paradise; but it is only the Septuagint translates it so; all the other Greek versions, and St. Jerome, the Vulgate, the Chaldee Paraphrase, and the Syriac, render it "from the beginning," or to that effect.

Warren concluded that kedem refers to the first land occupied by human beings and proposed translating *me-kedem* as "from the north." This does away with several discrepancies and harmonizes it with verses identifying the north as the abode of God (Job 23:9, 10, Psalm 48:2, Ezekiel: 1:4, 43:4). It also explains Genesis 11:2. Noah's descendants journeyed me-kedem into the plain of Shinar (Sumer) in the Tigris-Euphrates Valley, not "from the east," but from Mount Ararat in the north. By translating Genesis 29:1 as "children of the north" instead of "the east," we have the correct location for "the land of Uz" (Job 1:1). If *kol bnei kedem* in Job 1:3

was translated "He was the greatest of the sons of the north," it would align with Josephus, who wrote, "Uz founded Trachonitis and Damascus; this country lies between Palestine and Coelsyria."[10]

Moreover, if we translate me-kedem as "from the north" in Ezekiel 43:2, we have: "There, the glory of the God of Israel was coming [from the north]; the sound was like the sound of mighty waters; and the earth shone with his glory." It harmonizes with Job 37:22. "Out of the North He comes in golden splendor; God comes in awesome majesty." Vail concurs with Warren in identifying kedem as "the beginning spot," noting the error is also found in *The Egyptian Book of the Dead*.[11]

Many primeval words contain the archaic root KDM or KD. In Hebrew, Chet Kadmon is "Original Sin." Nachash HaKadmoni is the primeval serpent. Plato taught in the Grove of Akademos, the origin of the word *academy*. The Greeks viewed the Arcadians as their ancestors as the Babylonians viewed the Akkadians. Akkad is the circumpolar motherland, after which the Akkad of the Tigris-Euphrates Valley was named. The Babylonians located it on the "Mount of the North."[12] Kadosh in Hebrew means "pure, holy, sacred, consecrated." Kodkod means "top or crown." The related Greek word Kudos is "glory, fame." Since Eden is the place of the fall, the root KD is the parent of the Latin *cadere* (to fall) as in "cadence, cascade, cadaver," or something that befalls you like an "ac*cid*ent."

The Egyptian word *khet* is equivalent to the Hebrew word and letter *chet*.[13] Chet means, "sin, fence, broken and terrified." It represents the number eight, which stands for eternity and new beginnings. A primary meaning of *khet* in Egyptian is "steps which held up the Judgment Seat of Osirus." *Khet aa* is "the Great Throne of Osirus," who is often pictured seated on a seven-step staircase. Khet also means, "to go or sail to the North."[14] The determinative sign of a broken dead tree indicates the loss of the Tree of Life and the awful state of being closed off from the garden. Doubled, *khetkhet* means "to cut to pieces, destroy, break a command." Its Hebrew equivalent *chatat* means "be dismayed, crushed."

The following are definitions of *khet* and the related Hebrew words:

- khet ⸾ wood, tree, staff, scepter
- Khet en Ankh (Tree of Life)
- khet ⊜ pain, misery, anguish
- Hebrew *chat*—be broken, terrified
- khetem ⸾⸾⸾⸾ seal up, close, shut up, end

- Hebrew *chetem*—shut, close, seal
- khetemt 𓏃 treasures of the god
- Hebrew *chatan*—bridegroom
- khetem 𓋴𓏏𓊪 stamped bread[15]
- Hebrew *chatam*—hide, reveal a secret

The Roots of Eden

Delight yourself in the Lord and He will give you
the desires of your heart. —Psalm 37:4

What does the name Eden mean? The name is probably best translated as "a well-watered plain of delight." It is close to the Sumerian *edin* (plain, field), *edu* (cosmic waters), and *id* (river). The Akkadian word *edu* means "river-flood." The related Hebrew word *ed* is "the Primordial River."[16] It also means "mist, vapor, an upsurge of water." An *eddy* is a circular current of water. Related are the Greek *hyd* (water), the Latin *udus* (wet, moist), and the English *wet* and *wade* via the Indo-European root WED, from which we also derive *wedding*.

Eden beginning with an aleph אדן means "foundation, basis, pedestal, and the socket of a column." As the central park of the New Jerusalem, it certainly qualifies. The root of Eden begins with the eye/fountain letter ayin ע. The word *ed* עד means "eternity, witness, proof." Again, Eden qualifies. According to any Hebrew dictionary, Eden עדן literally means "delight, pleasure" as in the similar-sounding Greek *hedone* (delight) and *hedys* (sweetness, pleasure).

The same root is in Ida, the sacred mountain of Crete and Phrygia. According to Plato, the Trojans claimed descent from Ida,

> Ilium was built when they had descended from the mountain,
> in a large and fair plain, on a sort of low hill, watered by
> many rivers descending from Ida. ("Laws" trans. Benjamin
> Jowett)

Virgil called Ida "our race's cradle."[17] The Greek word *ideal* stems from memories of Eden's perfection, as does the Sanskrit *idavat* (refreshment), *ida* (praise), and *idana* (the act of praising). Related also is the Latin *ides* (middle and light).

In the center of the Norse Asgard is the stately green Idavoll (Plain of Ida). Known as "the plain that renews itself," it floated far above earth.[18] Four streams flowed from its central well, Hvergelmir (roaring cauldron). A river, Ifing, whose waters never froze surrounded it.[19] *Ida* is an old Teutonic word meaning "happy." Here, it was said, the gods "became wrinkled and gray" when they could no longer eat "the golden apples of Idun," which were in a *box.*[20]

The Egyptian word for father, At $\langle \frown$, begins with the feather of truth and also means "womb, house, god, king, prince, dew, moisture."[21] *At* is the root of Aten ⊗𝄐 (fertile ground, earth, land, estate, beauty). The Aten is "immovable" and "sevenfold."[22] All Egyptologists acknowledge it as celestial, but it was only present *before the Flood,* which the Egyptians called the primeval time,

> O thou Aten, who had thine existence in the primeval time.
> —"Hymn to the Aten," EAW Budge, *The Gods of the Egyptians* volume 2

T is the Egyptian D. Aten is the Egyptian Fatherland Eden.[23] To translate it as "sun-disk" is a huge error. The Egyptians *never* defined it that way, and its determinative signs do not resemble the present sun in any way.

The old Mandarin Chinese word *yüan* 园 (Cantonese *yun*) means "garden, origin, source, orchard, park, spring, and a high level plain." *Yüan* can also mean, "round, circular, a large jade ring, distant, and remove."[24] The old Chinese brass coin—the cash—has a celestial shape. Traditionally, it represented "the marriage of heaven and earth that produced all things," "the perfect man," and "the internal rectitude" of government leaders. Long used as an amulet of protection, today it simply means "good luck."[25]

The Cash 冷玉(ES)

Fuxi 父系 and Nuwa 女娲 are the Chinese originators of marriage, which took place on a mountain in the garden of Kunlun, which is the same as伊甸园 *yīdiànyuán* (Garden of Eden). They are brother and sister and son and daughter of the holy First Ancestor Shangdi 上帝. They are credited with the invention of writing and helping to order and civilize the world. They are pictured holding aloft a carpenter's square and a compass; their lower parts are entwined.[26] Another name for Adam is Ya Dang 亞當. The sign of God (Ya) is the first character.

In the sacred texts of Japan, souls of high rank ascend to Takama-no-Hara 高天原 (Plain of High Heaven), the dwelling place of the Kami (gods).[27] The principal stock of the people is Yamato. The records of Shinto were compiled in the eighth century for the purpose of confirming their "celestial origin."[28] Yamato is the hero of sumo wrestling. A straw rope marking the territory of the gods surrounds the ring, and purifying salt is thrown in.[29] The match is won by forcing your opponent out of the ring or forcing him to touch the ground with any part of his body other than the bottom of his feet. Yamato is the deity of the crossing who went down in a terrible storm.[29] Many tribal founders, including Romulus, Numa, Quetzalcoatl, Thor, and so forth, disappeared in a horrendous thunderstorm.

The Forbidden Tree

> And the Lord God commanded the man, saying, "Of every tree of the garden you may freely eat; but of the tree of the knowledge of good and evil you shall not eat, for in the day that you eat of it you shall surely die." —Genesis 2:16, 17

In both hemispheres, tales of Eden are so similar they had to have a common source. While differing in small details, all tribal accounts cite the first man and woman's disobedience as the cause of their expulsion. The people of Nias, west of Sumatra, say death came into the world because of eating forbidden bananas.[31] The Yuracaré of Bolivia assert that we all die because a primeval twin (Eve) ate from the *mani* plant, a South American peanut.[32] The Kachins of Burma blame their first parents, *H Kun Lu* and *H Kun Lai*, for eating the rice that lost them their blissful abode. Unable to return, they now had to work for a living.[33]

In their former home, the garden of sunshine, the Polynesian first ancestress plucked the forbidden *ape* plant and enticed her brother. The ape resembles *taro*, believed to have supernatural powers.[34] In Toltec tradition, the first woman disobediently picks roses from a tree and injures and disgraces herself and all her descendants.[35] The Efe people of Zaire say their supreme being made a man, Baatsi, out of clay, covered him with skin, and filled him with blood. He also made a woman and commanded them to have children. He gave them one rule: "do not eat of the tahu tree."[36]

Adam and Eve were seen as gods, because their Creator Father breathed His own life into them. As married siblings, they were imitated in the brother-sister marriages of Egyptian rulers. Their special creation as perfect adults inspired the ancestor worship that dominates all early cultures and continues today. Everything in ritual involves the ancestors. In the *Mahabharata,* the gods and ancestors are one and the same.[37] The Sumerian Adam, Tagtug was a god, a garment maker, and a metalworker. Educated by Enki/Ea:

> Joyfully he imparted to him his counsel. Tagtug he educated, he ... him and ... him. Enki fixed the fate of the plant(s) and placed (them) in the midst (of the garden) forbidding Tagtug to take from the Tree to eat. The Tree of life until he dies he will not see. (Stephen Langdon, "Semitic Mythology," *Mythology of All Races*, Vol. 5 1916)

A Change of Skin

> Then the eyes of both of them were opened, and they knew
> that they *were* naked; and they sewed fig leaves together
> and made themselves coverings. —Genesis 3:7

It is commonly held that Adam and Eve's outward appearance changed after they disobeyed God. They covered themselves in fig leaves and hid from Him, having lost their former luminous appearance. In place of their former garments of light, God clothed them in "garments of skin" (Genesis 3:21). This tale of a change of skin appears in many cultures. One common version found on all continents is that men could once change their skin like snakes, but they lost their immortality because the old skin was injured or ruined in some way.

Descent and Emergence Tales

> But after I uproot them, I will again have compassion and will bring each of them back to his own inheritance and his own country.
> —Jeremiah 12:15

After the fall, Adam and Eve were placed on a small satellite of the Earth of God in an environment that was, although imperfect, vastly superior to what we have now. Consigned to live in exile here, they saw their ancestral land shining in bright splendor above their heads. The Father meant this proximity to be both a blessing and a reminder of the time they had walked with Him in His perfect garden.

The descent of the first human beings from the sky world is a prime theme of world mythology. We also find many examples of the alternate version that our first ancestors emerged from the earth. Of course, we now know which "earth" they are talking about, so the two versions are describing the same event. Words for paradise, Eden, the netherworld, and Heaven are interchangeable. Here the weather is always perfect, and no one ages.

Persia

According to the ancient Persians, all people are descended from Mashye and Mashyane who once lived in the square Khwanirath:

> Ohrmazd (Ahura Mazda) spoke to Mashye and Mashyane thus: "You are man, you are the ancestry of the world, and you are created perfect in devotion by me; perform devotedly the duty of the law, think good thoughts, speak good words, do good deeds, and worship no demons!' Both of them first thought this, that one of them should please the other."
> (Bundahish Creation, chapter 6)

They lost their purity when they believed the devil Ahriman's lie.

Greece

In the familiar but late Greek version of the fall, Zeus formed for the first man the beautiful, delightful (but evil) Pandora, a name meaning "all-gifted," for his wife. She was given charge of a forbidden "box," or cube-shaped jar, which she opens, unleashing the plagues of labor, fatigue, sorrow, illness, and old age. All that remained was hope.[38] The Chinese have a traditional green cube-shaped jar called the Zong, which represents the earth.

America

The Hopi assert that we are only here temporarily. We were moved here from another world because of the failure to adhere to the Creator's plan.[39] The Iroquois say a heaven above, where sorrow and death were unknown, was the prototype of this world and the setting for the first act of a cosmic drama. A "Heaven-tree" grew beside the lodge of the Great-Chief-He-Holds-the-Earth.[40] According to the Seneca, the people formerly lived in the sky with the Great Chief. When the chief's daughter fell sick, a wise man placed her next to a tree, which he dug up. Another man, angry about the destruction of the tree, kicked the girl into the hole. Then "Star Woman floated down through space."[41]

According to the Chorote of Gran Chaco, the first man and woman were created in the womb of the morning star.[42] The Arawak, the Yekuana, and the Yanomamo of Guiana say they descended from the sky world.[43] The Caribs claim descent from Louquo:

> Louquo, the first man, came down from the sky and other men were born from his body. After his death he ascended into the heavens. (HB Alexander, "Latin American Mythology," Vol. 11 *Mythology of All Races* 1916)

According to the sixteenth-century Inca historian Garcilaso de la Vega, this culture believed that "Our Father" sent down his first son and daughter from heaven. Manco-Capac and Ocllo emerged in Lake Titicaca (Island of the Sun) from their home in the sky. They were given a golden rod and were told that wherever it entered the earth, they were to establish themselves. The rod pierced the ground at Cuzco (navel). Manco-Capac declared himself "king of the four corners of the earth."[44] Titicaca was once 12,500 feet lower, and marine crustaceans and mollusks have been found in the lake.[45]

The Aztec couple Cipactonal and Oxomoco lived before the Flood in a state of purity and blessedness in Tamoanchan (House of Descending or Birth), the abode of Tonacateuctli (Lord of Sustenance), the Creator of the earth. In the center, amidst flowing streams and colorful birds, stood a tree whose branches and blossoms they were forbidden to touch. All was well until they broke off some of its blossoming boughs. For this disobedience, they were cast out. All men and women are descended from this pair.[46]

The Tempter (Aztec Codex Borbonicus)

In another Aztec account, the Earth Mother Xochiquetzal (Precious Flower), bride of the sky, lived on a mountain almost as high as the moon. She was the first to sin by eating of the sacred tree. The tree withered, broke in two, and she fell from the sky. In the Chichimec version, Itzpapalotl (Obsidian Knife Butterfly) is the mother who destroyed her children. Depicted with wings and a de-fleshed face, she has talons for hands and feet and wears a skirt of knives. She is shown in the codices with "the broken and bleeding tree of paradise." She was hurled down to earth for disobeying the divine command. All these personages are associated with the north.[47]

In the Creek migration legend, the first people came forth from the earth, crossed a river of blood, and came to a "singing mountain" where they received "mysteries and laws."[46] From this "hole of emergence," the waters of the Flood welled up.[47] The Jicarilla, an Apache group, say they emerged from "a great swelling womb," entering through an opening on the top of a mountain after the waters of the earth were broken.[48] The Algonquin, Iroquois, and Huron recall their origin in a higher world above the clouds.[49]

The Inuit of the Bering Strait relate that a long, long time ago, a man and woman came down from the sky and landed on one of the Diomede

Islands.[50] The Cheyenne say their ancestors came from the far north, beyond a great body of water:

> In the beginning, the Great Medicine created the earth and the heavenly bodies; and, in the far North, a beautiful country, an earthly paradise where fruits and game were plentiful, and where winter is unknown. Here the first people lived on honey and fruits: they were naked and wandered about like the animals with whom they were friends; they were never cold or hungry.[51]

The Algonquins say their ancestor Ataensic, trying to cure her sick husband, was cast out of heaven for cutting down a tree with curative powers. The tree fell over a precipice, and she fell in after it.[52] A Black Foot girl was cast out of heaven, the morning star, for digging the forbidden turnip. As punishment, she and her son were wrapped in an elk skin and lowered to earth.[53] The Dakotas describe the descent of "the Woman-from-Heaven" in the north.[54]

Ireland

Though you have made me see troubles, many and bitter, you will restore my life again; from the depths of the earth you will again bring me up.
—Psalm 71:20

The Leabhar Gabhala Eireann (*The Book of the Taking of Ireland*, historical poems and prose compiled in the eleventh century) traces the Irish people from Adam to Noah to Japheth. They also had a cosmic history paralleling the events of Genesis 6, the invasion of the sons of God. This is also the case with five figures in the *Rg Veda*.[56] Five brothers descended from their huge deformed mother Danu. Even though she had only one eye, one arm, and one leg, she possessed the strength of the entire tribe. All of these brothers battled the Fomorians, a monstrous race that descended from Ham or Cain, whose king, Balor, had one fearsome eye.[57]

The last to arrive were the Tuatha Dé Danaan (divine tribe), a parallel of the Welsh *Don*, from the north. They were *ande* (giant gods) who came through a fog from the sky, landing on a mountain. Their king, Nuadhu, had one silver arm, but Balor, king of the Fomorians, killed him with his evil eye.

Lugh (in Wales Lleu Llaw Gyffes), who succeeded Nuadhu, killed Balor with a slingshot in the eye. He was a singing god, the inventor, and master of all arts and skills, including warrior, healer, poet, and blacksmith. He had one arm, one leg, one eye, and magical powers. Eventually defeated by the sons of Mil from Spain, the Tuatha retreated to the *sidhe* (pronounced "shee"), the ancestral hill of the dead.[58]

Africa

> But God will redeem my life from the grave;
> he will surely take me to Himself.
> —Psalm 49:15

Belief in a supreme being is thoroughly African in origin and existed long before any Moslem or Christian influence. He created everything and knows and sees all. He is respected as a kind and merciful spirit who cares for the people and does not strike them with terror. He personifies justice, rewarding good and punishing evil, and it is to His final court that even the poorest can appeal. Throughout the continent, He is given names and titles like "Giver of Breath and Souls, the Bow in the Sky, the One Who Bends Even Kings, the One You Meet Everywhere, Greatest of Friends, God of Pity and Compassion, the Inexplicable, and the One Beyond All Thanks."[59]

Tales of the origin of death and God leaving in anger are common throughout Africa. A woman is usually blamed. The Dinka of eastern Sudan assert that a wall in heaven held human beings in the sky, but people ate part of the wall, and God pushed them down to earth. The Yoruba of Nigeria say the first people were made in heaven and sent to earth by the Great God. First man preached morality and family order.[60] The ancestors of the Molama clan (Zulu) were a man and woman who came down from the sky and alighted on a certain hill. The Ngombe of Zaire say no one lived on earth in the beginning. The first people lived happily in the sky with Akongo (God).[61]

The Shilluk of Sudan say the people lived in the land of God at first, but God sent them away for eating fruit that made them sick.[62] According to the Barotse of Zambia, the Creator Nyambi drove his special creation Kamonu, the first man, out of his sacred realm, Litoma. He gave Kamonu a garden to tend, hoping to keep him out of mischief, but Kamonu engaged in murderous behavior, and Nyambi decided to move away from the earth.[63] The Mossi of Sudan assert that the earth god is angered upon seeing the shedding of human blood.[64]

Pacific Islands

Set your minds on things that are above, not on things that are on earth.
—Colossians 3:2

Traditions from New Guinea, the Kei Islands, Indonesia, the Marquesas, and the Carolines have a common tradition that their ancestors were a couple who came down to earth from heaven. Hina, the first woman, a goddess, brought both life and death to mankind.[65] The Ifugao of Kiangan in the Philippines claim that the first son of Wigan, Kabigat, came from the sky region, Hudog, to the earth world. In the Carolines, it is said that Lingobund descended from the sky to the earth and gave birth to three children who became the ancestors of humanity.[66]

The Maori of New Zealand say that the deity Tiki (or Tané) molded the first man from his own blood mixed with red clay. He animated the figure by breathing into him his own breath. The man looked so much like himself that he called him Tiki-ahua (Tiki's likeness). He next meditated how he could make a companion for him, so he took a rib from the man, and the first woman was produced. Then he ordered her to live with Tiki-ahua as his wife, and by them, the whole world was peopled.[67]

When missionaries first arrived in Hawaii in 1820, they asked the natives about their origin. They were told that they descended from *reva* (sky, heaven, the Kingdom of God). The words used did not refer to an ocean voyage but a movement through space. This knowledge of divine origin was preserved in temple chants and prayers. The Creator god Tané (or Kané) had originally placed the first man, Kumuhonua, in a beautiful garden, Kalana-i-hauola. He produced a wife for him called Ke-ola-Ku-honua from his right side. In the midst of the garden was a holy apple tree, whose fruit caused the death of strangers.[68]

The Names Adam and Eve

And the Lord God formed man of the dust of the ground, and breathed into his nostrils the breath of life; and man became a living being.
—Genesis 2:7

The names Adam and Eve can still be found in some tribal records. The word "'breath" is *atem* in German and *adem* in Dutch. In Sanskrit, *adima* means "first, primitive, original." It's not surprising that the first man is honored in place names, such as Adama in Ethiopia and the Adamawa Mountains in Nigeria. Some have lost the first "a," like Damascus in Syria and Damas in Egypt. Adamhi in Jordan became Damiya.[69]

The name Hava (or Eva) is widespread in words for life, separation, and Heaven itself. Havdala is a Hebrew ceremony at the end of the Sabbath separating it from the weekdays. The similar Mayan word *eua* also means "separation." In Sanscrit, Eva means, "strengthening an idea and earth." We also find the "eve" root in *jiva* (breath, soul, life), *diva* ("heaven, sky"), *deva* ("divine") and *ziva* ("happiness"). It is the life force in the Persian *hvov*, the Hitite and Aramaean *Hawwah*, and the Latin *avus* (ancestor).

In Tahiti, it is said that the chief deity Taaroa made the first man out of red earth. Later, he put him to sleep and took out one of his bones (ivi) and made a woman. She became his wife, and they became the progenitors of all people. Ivi is a native word meaning "bone, widow and a victim slain in war."[70] Similar traditions are found in other parts of Polynesia. A rib was taken from first man's left side and "started up a live woman." The man called her Ivi (rib). To the Mbyá, a Brazilian tribe, the term *iva* means "paradise" or "the sky, the present home of the Father." *Ivi mará ei* means, "land without evil."[71]

After the Fall

> From one man he made every nation of men, that they should
> inhabit the whole earth; and he determined the times set
> for them and the exact places where they should live.
> —Acts 17:26

Our recorded histories go back thousands, not millions, of years. Highly advanced civilizations emerged fully developed. All of the elements upon which the civilizations of later ages would be based—complex hybrid grains, advanced architectural techniques, functional pottery, even the beginnings of metalwork—developed almost simultaneously. The beginning of writing grew up virtually overnight with the sudden appearance of sacred symbols and complex language. No evolution from simple to complex is seen. There are *no* fossil Paleolithic human cultures![72]

In her book *Plato Prehistorian*, Mary Settegast explains how extraordinarily advanced communities emerged out of nowhere:

> Something is fundamentally wrong with the way we have pictured Paleolithic man and, by extension, the course of human prehistory ... Virtually all of the skills that would form the basis of the civilizations of later times were suddenly present - advanced architectural and interior designs, fully domesticated animals and complex forms of grain, even a functional pottery tradition and the earliest signs of metal work. (Mary Settegast, *Plato Prehistorian*)

The Father wanted His children to share his home forever, but sin changed all that. The expulsion from Eden was no punishment but an act of mercy. A loving Father cut off access to the Tree of Life to save them, for eating its fruit would have made them immortal in a state of spiritual death, separating them from Him forever (Psalm 145:17). Sinful humans cannot survive in the presence of a holy God (Exodus 33:20).

One of the greatest blessings in knowing about a celestial Eden is that it forever does away with the notion that the human race is related to any creatures previously existing here. Why would the Creator plant His perfect garden on a world with a long history of death and destruction written into its very stones? We have an immortal soul, a sense of someone far greater than ourselves, and the desire and ability to communicate with Him. Mircea Eliade defines this as the nostalgia for paradise, "what a Christian would call the state of man before the fall."[73]

The Father prepared this planet for human habitation by wiping out species incompatible with our survival. He then transferred Adam and Eve to this flawed but beautiful planet we now call home. The fall was physical as well as spiritual. Charles Darwin and his followers have never been able to explain the three Cs—consciousness, cognition, and conscience—and what evolutionary advantage is there in singing in perfect four-part harmony and playing instruments?

Aliens did not put us here, my friends. "The fool has said in his heart, 'there is no God.' They are corrupt, they have done abominable works there is none who does good" (Psalm 14:1). But God is merciful and full of loving-kindness. "I tell you that in the same way there will be more rejoicing in heaven over one sinner who repents than over ninety-nine righteous persons who do not need to repent" (Luke 15:7).

After the fall, Adam and Eve longed to be reunited with their heavenly Father. As they gazed up at their lost homeland with tears and anguish, they drew pictures of it on rocks and clay. This spawned the development of writing. After the Flood, the survivors could no longer see their homeland, but the sacred symbols lived on. The Targum Yerushalmi of Genesis 2 states that to form Adam, God took dust from the four quarters and from the center of the original Eretz (earth).[74]

> Of man's first disobedience, and the fruit
> Of that forbidden tree, whose mortal taste
> Brought death into the world, and all our woe,
> With loss of Eden, till one greater Man
> Restore us, and regain the blissful seat.
> (John Milton, *Paradise Lost*)

Chapter 10

The Island of God

Where Was Atlantis?

For there is nothing hidden that shall not be revealed.
—Matthew 10:26

When Plato wrote of Atlantis sinking beneath the ocean in a great cataclysm, he started a controversy lasting to the present. His dialogues of Timaeus and Critias, written around 360 BC, seemed to be the only source of the legend. To some historians, the extreme suddenness of civilization's emergence makes the former existence of Atlantis plausible. Plato obtained his facts from the priest Solon, the great lawgiver of Athens, who heard it from the Egyptians. He described it as an island larger than Libya and Asia together, an ordered and lush paradise.

Where was Atlantis? To be taken seriously, any theory must be consistent with all the details provided by Plato. Explorations of the Atlantic Ocean floor, or any other sea floor, have revealed no sunken continent. Francis Bacon thought that Atlantis was America, that it was said to have disappeared because human beings lost the knack for getting there.[1] Others believed it to be the island of Thera in the Mediterranean Sea, destroyed in a massive volcanic eruption around 1627 BC–1600 BC. None of these theories fit Plato's description.

In the Sumerian *Epic of Gilgamesh*, paradise is said to be an island encircled by the waters of death.[2] Souls had to cross those waters in order to enter the netherworld. The ancients had many island paradises, among them Ogygia, Aeaea, Avalon, Delos, Thule, and Homer's floating isle of Aeolia. Descriptions are remarkably similar. All were in the north, and all disappeared, including that ancient favorite of all islands, Hyperborea.

Star Islands

The sky vanished like a scroll rolling itself up, and every
mountain and island was removed from its place.
—Revelation 6:14

Ancient historians wrote of the happy Hyperboreans, a Greek term meaning "those beyond the North wind," or "those above the mountain." Said to be virtuous vegetarians, they were a sacred race with a lifespan of a thousand years.[3] Greek historian Diodorus Siculus (first century BC) said that Hyperborea was as big as Sicily and bordered by the earth-encircling river Okeanos. It was productive of every crop. In the center was the temple of Apollo, to whom the Hyperboreans daily sang praise,

> Furthermore, a city is there which is sacred to this god, and the majority of its inhabitants are players on the cithara; and these continually play on this instrument in the temple and sing hymns of praise to the god, glorifying his deeds: (Library of History 2. 47, trans. CH Oldfather)

The Greek poet Pindar described the difficulty of traveling there:

> Of the fairest glories that mortals may attain, to him is given to sail to the furthest bound. Yet neither ship nor marching feet may find the wondrous way to the gatherings of the Hyperborean people. (Pindar, Pythian Ode 10 trans. GS Conway)

The Roman historian Pliny the Elder wrote:

> A happy race of people called the Hyperboreans, who live to extreme old age and are famous for legendary marvels. Here are believed to be the *hinges on which the firmament turns*

and the extreme revolutions of the stars, with six months' daylight and a single day of the sun in retirement, not as the ignorant have said, from the spring equinox till autumn: for these people the sun rises once in the year, at midsummer, and sets once, at midwinter. It is a genial region, with a delightful climate and exempt from every harmful blast. The homes of the natives are the woods and groves; they worship the gods severally and in congregations; all discord and all sorrow is unknown … Those who locate them merely in a region having six months of daylight have recorded that they sow in the morning periods, reap at midday, pluck the fruit from the trees at sunset, and retire into caves for the night. Nor is it possible to doubt about this race, as so may authorities state that they regularly send the first fruits of their harvests to Delos as offerings to Apollo, whom they specially worship. (*Natural History* 3 trans. H Rackham)

The Greek poet and scholar Callimachus (310–240 BC) wrote that the Hyperboreans celebrated annual choir festivals.

Wherefore from that day thou art famed as the most holy of islands, nurse of Apollo's youth. On thee treads not Enyo nor Hades nor the horses of Ares; but every year tithes of first-fruits are sent to thee: to thee all cities lead up choirs, both those cities which have cast their lots toward the East and those toward the West and those in the South, and the peoples which have their homes above the northern shore, a very long-lived race. (Hymn 4)

The original inhabitants of Easter Island said they came from an enchanted "star island" of incredible beauty that no one could find. The land was crossed with roads leading out to all directions. These were so artistically paved there were no rough edges. Branches of coffee trees were laced together like muscles overhead, resembling a spider's web. The divine builder of these roads sat in the center in the place of honor.[4]

P'eng-lai 蓬莱山 are the Chinese Islands of the Blessed. They contain the palaces and mansions of the immortals. Here they drink from a jade fountain of life flowing from a jade stone one thousand *chang* high (approximately

3,580 meters or 2.2 miles). The shores of P'eng-lai are covered with gems, and the Ling Chih (Plant of Long Life) grows in its center.[5] These islands floated freely at first, until the immortals persuaded the emperor of heaven to anchor them. Later on, two of the islands drifted north and sank. The emperor became angry and reduced all giants in size.[6]

The Malay word for rainbow, Palangi (striped), resembles the name P'eng'lai. Before the introduction of Islam, the people worshipped Toh Batara Guru, "the all-powerful spirit who held the place of Allah ... a spirit so powerful that he could restore the dead to life; and to him all prayers were addressed."[7] In Malay tradition, the firmament was an island. Pauh Janggi is the universal Malay name for the tree that grows on that island in the central whirlpool Pusat Tasek (Navel of the Seas).[8]

The Duwa people of Arnhem Land (Australia) say their ancestors originally came from the island of Bralgu and return there after death. They say the morning star rests there on a tall pandanus palm tree. This is the dreaming tree of life and death.[9] The people still sing its ancient song. The Primordial Sky Being of the Kaitish tribe of central Australia expelled certain of his sons and daughters from the sky to the earth when they neglected his sacred services. These became the ancestors of all humanity.[10]

The Whirlwind of the Lord

> Behold, a whirlwind of the Lord has gone forth
> in fury—A violent whirlwind!
> —Jeremiah 23:19

Instead of thy making a cyclone,
Would that a lion had come and diminished mankind.
Instead of thy making a cyclone
Would that a wolf had come and diminished mankind.
Instead of thy making a cyclone
Would that a famine had arisen and [laid waste] the land.
Instead of thy making a cyclone
Would that Erra (the plague god) had risen up and [laid waste] the land.
—*The Epic of Gilgamesh,* third millennium BC, 11th Tablet

Tales of a total destruction of an island are told throughout Polynesia and Micronesia. Although differing on some minor points, all attribute the disaster to an angry god. The island of Tahiti was totally destroyed in a terrible hurricane in which the stones and the trees were whirled aloft then fell from the heavens. Water from the deep drowned every living creature, except a husband and wife, who were saved along with some chickens, a dog, a kitten, and a pig (the only domesticated animals known then). The couple took refuge on a high mountain, Mount O Pitohito. From these two people, all human beings are descended.[11]

The Assyrian tablet of Nineveh recalls the terror of those days and the coming of the catastrophic Abubu:

> Swiftly it mounted up ... [the water] reached to the mountains
> The god Shamash had appointed me a time (saying)
> The Power of Darkness will at eventide make a rain-flood to fall
> Then enter into the ship and shut thy door.
> The appointed time drew nigh;
> The Power of Darkness made a rain-flood to fall at eventide.
> I watched the coming of the [approaching] storm,
> When I saw it terror possessed me,
> I went into the ship and shut my door.
> To the pilot of the ship, Puzur-B?l (or Puzur-Amurri) the sailor
> I committed the great house (i.e. ship), together with the contents
> As soon as the gleam of dawn shone in the sky
> A black cloud from the foundation of heaven came up.
> Inside it the god Adad (Ramm?nu) thundered,
> The gods Nab? and Sharru (i.e., Marduk) went before,
> Marching as messengers over high land and plain,
> Irragal (Nergal) tore out the post of the ship,
> En-urta (Ninib) went on, he made the storm to descend.
> The Anunnaki brandished their torches,
> With their glare they lighted up the land.
> The whirlwind of Adad swept up to heaven.
> Every gleam of light was turned into darkness ...
> the land ... as if ... had laid it waste.
> A whole day long [the flood descended] ...
> [The water] attacked the people like a battle.

Brother saw not brother.
Men could not be known (or, recognized) in heaven.
The gods were terrified at the cyclone.
They took to flight and went up into the heaven of Anu.[12]

It is evident from our ancestral eyewitnesses, Noah and his family, that the Flood began when God's protective canopy over the earth became the greatest cyclone humanity has ever experienced. In India, it is called Mandara, the churning mountain of the gods in which so many precious things were lost. The Mahabharata describes a scenario where gods churned the ocean with Mt. Mandara. A terrible roar was heard as the sea creatures were crushed, and the friction from trees crashing into each other caused numerous fires, killing many land animals. Then Indra sent the great rain to quench the fire.[13]

The following Semitic words for cyclone and flood are very similar,

Hebrew—*mabul* (flood, storm)
Sumerian—*abuli* (cyclone)
Babylonian—*abubu* (cyclone)[7]

The Sumerian version of the Hebrew Yom Mabul (Day of the Flood) is Um Bubuli (the Day of the Snatching Away). This was a time of great sorrow, fasting, and mourning all over Mesopotamia.[14]

The Creator is associated with a whirlwind in a different way among the natives of northwest Australia. They claim to be the children of the all-merciful Creator Gnurker and that they sprang from one man and woman. When God saw this earth was fit for humans, with plentiful animals and fish, He sent them down with instructions on observing ceremonies and marriage laws by means of an immense *whirlwind* reaching from heaven to earth. They were to strictly follow His commands, and when they died, their spirits would be received in heaven. They had control over every living thing. By a ceremony of will, they could cause increase or decrease.[15]

---

[7] The word for bubonic plague (bubo) is related. The people believed it was divine judgment.

The Great White Throne

> Then I saw a great White throne and Him who was seated on it. Earth
> and sky fled from His presence, and there was no place for them.
> —Revelation 20:11

The isle of Zeus, the cradle of the human race, was called Alba (white) by the
early Latin tribes. Alba was the political capital of the Latin League until its
rival Rome wrested it away.[16] In the holy tongue, the Italic Alba means "God
the Father." Alba survives in the names Alps, Elbe, Albany, Albania, and
Mount Alburz, the Persian Mount Zion. Silvery whiteness is very common
in myths referring to the center of heaven.

The sacred king of folklore commonly departs to a revolving white or
silver island abundant with orchards to be restored to his former strength.
There he waits for his time for return. These island paradises include the
Welsh Avalon and the fair Isle of Emain of Goidelic lore, which departed in
a great cataclysm. The glass castles of Irish, Manx, and Welsh folktales are all
island shrines encircled by glassy-green water.[17]

The Greeks had several names for the White Island in the far north.
One was Omphalos Thalasses (Navel of the Sea), said to be so beautiful that
once seeing it, one was seized with wonder and delight.[18] Its central fountain
parted into four streams. Divided into quarters, Plato called these the Islands
of the Blessed:

> Now in the days of Cronos there existed a law respecting the
> destiny of man, which has always been, and still continues
> to be in Heaven - that he who has lived all his life in justice
> and holiness shall go, when he is dead, to the Islands of the
> Blessed, and dwell there in perfect happiness out of the reach
> of evil; but that he who has lived unjustly and impiously
> shall go to the house of vengeance and punishment, which
> is called Tartarus. (Plato, The Dialogues of Plato, "Gorgias"
> trans. Benjamin Jowett)

The Egyptians described the ivory palaces of the righteous on the beautiful
Isle of Yebu (Elephantine) ⌖. Yebu was the first city that ever existed. The
cavernous sources of the Nile were in its center.[19] Its earthly counterpart is an
island in Nubia. The magnificent white wall surrounding the sanctuary of

Ptah in the city of Memphis was modeled after the celestial Yebu. The name Memphis derives from Mennefer (beautiful heaven).[20]

To the indigenous peoples of the Housatonic Valley of New England, the guardian of a tribe is a white animal or bird, and the death of a white deer is said to bring misery to the tribe. The same animal causes disaster in Japanese mythology, and the hero Yamato becomes a white bird at his death.[21] The Dakotas revere a white buffalo and look for its symbolic rebirth. In European legend, there is a corresponding white doe or unicorn.[22] The Druid priests of the Celts sacrificed white bulls and venerated a white sow. The Persian Primeval Ox of the Eran Vej was brilliant white, shining like the moon.[23]

The legendary motherland of Polynesia is Ta Rua (also called Tahiti-Na, Havaii-ti, or Hava-Iki), an enormous island continent where life and civilization began. The Samoans call it Mu. Because the people had forsaken the commandments of Tané and had become evil, it was blown to bits by volcanoes and smashed by titanic tidal waves. Rent asunder by earthquakes, it sank beneath the sea.[24] Note the similarity of the name Ta Rua to the Egyptian word Ta (earth, land) and the Hebrew *ruah* (spirit).

Paradise Islands

The Lord is not slow in keeping his promise, as some understand
slowness. Instead he is patient with you, not wanting anyone
to perish, but everyone to come to repentance.
—2 Peter 3:9

The Dakotas say all Indians came from a primal island destroyed in a great catastrophe.[25] The Zuni First-world was a large island surrounded by an ocean. Their village, Itiwana (middle place), reflects the cosmos. Zuni clans are divided into seven groups.[26] Tribes of the Pacific Northwest tell of a lost isle in the middle of the ocean called Samah-tumi-whoo-lah (White Man's Island). A race of white giants lived there, ruled by the powerful Scomalt. For many years, they lived in peace, but quarreling began, which led to war. Scomalt grew angry, and only one man and woman escaped. This occurred "when the sun was no bigger than a star."[27]

Aztec man-made island of Tenochtitlan ⚠
Friedrich Peypus (1485–1534)

The Mexica, an Aztec tribe, said their ancestors, "the gods," emerged from "seven womb-caverns" on an island called Aztlan (surrounded by water), also known as Anáhuac (the Place within the Ring). This island paradise of dazzling whiteness had a white cypress tree from which issued a fountain. White willows, white reeds, and bulrushes surrounded it. This promised land was found by searching for a lake with a cactus in the center on which perched a white eagle.[28] It was here that they built Tenochtitlan, now the site of Mexico City.

Another island modeled on an island paradise is off the coast of Santiago Ixuintla in the Mexican state of Nayarit. Here is the amazing pre-Columbian man-made city-island of Mexcaltitán de Uribe looking very much like the Land of God:

The Aztecs also patterned it after Aztlan, their celestial homeland.

In their *Annals,* the Cakchiquels of Guatemala say their ancestors came from the fourfold Isle of Tulan on a white mountain. From there, we were begotten at the zenith where "Our maker, the Creator" dwells. When passing through the gate of Tulan, our ancestors brought a red staff, which they thrust into the waters. These divided, and they passed over. The Cakchiquels have seven tribes, seven rulers, seven villages, and seven nations.[29] Before men planted maize, they refrained from sexual relations and burned incense at the four corners of their fields, which were large and productive. Firstfruits were consecrated for holy use.[30]

According to the ancient records of Shinto, there was chaos in the beginning, and the Eternal-Ruling-Lord with two deities, the original triad of generation, produced "all gods, men, and things." Later on, by divine command, the married deities, Izanami and Izanagi, descended from their original home by means of "the Floating Bridge of Heaven." They landed on the island Onogorojima (Island of the Congealed Drop).[31]

The original inhabitants of Japan, the Aino, who called themselves "offspring of the center," claimed descent from the north and bury their dead facing that direction. In their language, Yezo is the Island of Heaven, Yasu means "peace," and Ame-No-Yasu-No-Kapa is "Heavenly Tranquil River."[32] According to traditions of mainland Southeast Asia, the first couple lived on an island, but they left the island and reached the mainland. Then a huge catastrophe cut them off from their parent land forever.[33]

Asgard, Odin's glorious city, is set high above the heavens upon a holy island whose beauty is unequaled in nine worlds. In the middle of Asgard is Idavoll, the Court of Judgment, the High Thingstead of the gods, and Odin's golden throne. A broad river of fierce currents and leaping flames surrounds it and a lofty wall. It had only one entrance, Odin's mighty gate. No one unworthy could enter in.[34]

Atlantis, the Antediluvian World

For evil men will be cut off, but those who hope
in the Lord will inherit the land.
—Psalm 37:9

In his book, *Atlantis, the Antediluvian World,* former Minnesota Congressman Ignatius Donnelly identified Atlantis as the seat of all races, the location of

the four rivers, and source of the universal sign of the cross.[35] He noticed
that biblical descriptions of the Antediluvians and Plato's descriptions of the
Atlanteans have a great deal in common.

- Both the antediluvian world and Atlantis were identified with human
  origins.
- Both were described as island paradises.
- Both the Antediluvians and the Atlanteans had a population
  explosion.
- Both lived happily and peacefully, having everything provided for
  them.
- Both were degraded by intermarriage between divine and human
  stock.
- Both degenerated and were punished by God for their great
  wickedness.
- Both were destroyed by water.

Warren also concluded that Atlantis was the antediluvian world.[36]
Donnelly and Warren were right about many things, but they did not
differentiate between heaven and earth, so their Atlantis, Eden, Mount
Olympus, Tulan, Aztlan, and so on were all located on this planet. This
is completely understandable since the two earths were once connected by
Mount Zion and their traditions were intricately intertwined.

In his unfinished dialogue "Critias," Plato described the moral degeneration
of the Atlanteans. Compare his description with the Antediluvians. Plato
noted that as long as they had "the divine nature," they revered and obeyed the
gods and were virtuous to one another, not pursuing gold and luxury. When
they mixed with mortal stock, the divine nature in them weakened, and they
degenerated into the pursuit of power and ambition:

> Zeus, the god of gods, who rules according to law, and is able
> to see into such things, perceiving that an honourable race
> was in a woeful plight, and wanting to inflict punishment on
> them, that they might be chastened and improve, collected
> all the gods into their most holy habitation, which, being
> placed in the centre of the world, beholds all created things.
> And when he had called them together, he spake as follows—
> [here his account ends]. ("Critias" trans. Benjamin Jowett)

The belief that the First Land was an island that sank beneath the sea resulted in the Creator being perceived as a sea god (e.g., Italic Neptune and Irish Manannán mac lir). Plato identified the god of Atlantis as Poseidon, the pre-Hellenic god of the sea who received the island in allotment. The name means "Lord of the Earth" or "Husband of Earth." Homer calls him "encircler of the earth."[37] His epithets were Prokystios (the flooder) and Enosichthon (earth shaker). All earthquakes were his work. Poseidon was known for his gigantic children who lived beyond the Pillars of Hercules. He was said to have fought in a battle against the giants. His altar stood before the Erechtheum on the Acropolis.[38]

The Mystery Solved

With His mighty wind He will shake His fist over the River, and strike it in the *seven streams,* and make men cross over dry-shod.
—Isaiah 11:15b (NKJV)

The ancients associated Atlantis with the center and the pole. Euripides called the pole Atlantean:

"Twin bears, with the swift-wandering rushings of their tails, guard the Atlantean pole." (*Star Names: Their Lore and Meaning*, Richard H. Allen 1889)

Plato described the crossed streams of Atlantis, one warm one cold:

He (Poseidon) himself, being a god, found no difficulty in making special arrangements for the centre island, bringing up two springs of water from beneath the earth, one of warm water and the other of cold, and making every variety of food to spring up abundantly from the soil. ("Critias" trans. Benjamin Jowett)

Like the Kingdom of God, Atlantis was associated with pillars. Solon located Atlantis beyond or within the Pillars of Hercules. They are also called the Pillars of Kronos. In the *Voyage of Bran,* the Welsh Avalon is described as an island of delights with four golden pillars supporting it. There was no pain, sickness, sorrow, or death.[39] Atlantis was also associated with a wheel. The Isle of Delos (appearing) was the center of the Cyclades (wheel, cycling) in the north. Floating at first, Apollo fixed it there.[40] Herodotus said that the natives, called "Atlantes," took their name from it.[41] In a parallel tradition, the isle of Rhodes was spun on a golden spindle at the prayer of Helios.[42]

Seven concentric circles of land and water enclosed Atlantis:

> Poseidon created alternate zones of sea and land larger and smaller, encircling one another; there were two of land and three of water, which he turned as with a lathe, each having its circumference equidistant every way from the centre, so that no man could get to the island, for ships and voyages were not as yet. (Plato "Critias" trans. Benjamin Jowett)

India had two very ancient names: Bharata and Jambu-dwipa. Dwipa means island, and Jambu was a delightful central plateau surrounded by seven concentric islands alternating with seven seas increasing in height toward the center.[43] The Vishnu and Vayu Puranas describe Mount Meru in the center of Jambu-dwipa. It had four enormous pillars—gold, silver, brass, and iron—and four rivers flowed from it. Jambu is also the name of the Tree of Life, said to be a myrtle, a pipal, or an apple tree with fruits as large as an elephant.[44]

The Irish paradise of Tir na n-Og is likewise surrounded by seven zones of land and seven of water. The Fomorians called it "the Island of the Living" or Breasul.[45] It is described in the Battle of Gabhra:

> Tir na n-Og is the most beautiful country that can be found,
> The most productive beneath the sun;
> The trees are bending under fruit and bloom,
> While foliage grows to the top of every bramble,
> Wine and honey are abundant in it.
> And everything the eye ever beheld;
> Consumption shall not waste you during life,
> Neither shall you see death or dissolution.
> (Trans. N O'Kearney)[46]

Donnelly correctly made the connection of Atlantis to Eden:

In the great ditch surrounding the whole land like a circle, and into which streams flowed down from the mountains, we probably see the original of the four rivers of paradise, and the emblem of the cross surrounded by a circle, which as we will show hereafter, was from the earliest pre-Christian ages, accepted as the emblem of the Garden of Eden.[47]

The compass ⊕ was the emblem he referred to.

In summary, the ancient Island of Paradise was:
1. The First Land
2. The original earth and cradle of the human race
3. Located in the far north at the pole
4. A silver or white island on a white mountain
5. A land with crossed streams or pillars facing the four directions
6. Surrounded by seven concentric circles
7. Lost under the waves in a great cataclysm

No better proof exists than the ancient cross of Atlantis:

Cross of Atlantis (ES)
Linguistic footnote

Atlantis begins with the archaic roots AT and TL. In Egyptian, AT means "father." ATA means "dew or moisture." The root TL (or DL) also has water, mountain, and paradise connections.

Hebrew—*tel*—dew, moisture, hill, mound

talal—cover with beams
tallit—covering prayer shawl
taltela—hurling down

Akkadian—*tul-ku*—illustrious mound, holy altar

Mayan—*tulel*—soul
Nahuatl—*atl*—water
atlatl—spear thrower

Indo-European: TEL—to lift, bear up   TEL—to aim   DEL—to drip

Greek—*tholos*—vaulted chamber, heaven
telos—turning point,
tele—far off
stalaktos—dripping
thalassa—fountain or sea

Sanskrit—*atala*—underworld, hell, bottomless
tula—a balance, Libra
tal—to fix in one place, establish
tala—base, level, palm of hand, sole of foot
talaka—a pond, fragrant earth, clay pot
tali—Hindu marriage symbol

English—tallow
tolerate
deluge
delta—mouth of a river
dell or dale (river valley)

Gaelic—*tulca*—large wave, heavy rain,
tuile—flood

The letter D (originally △) Hebrew Dalet (door), Greek Delta, is a picture of the mountain/canopy.

Sacred Places

Atlantis—Plato's sunken continent

Aztlan—Aztec celestial homeland

Dal—a boundary stone

Delos—island with an oracle (mouth), legendary birthplace of Artemis and Apollo

Delphi—oracle of Apollo

Dilmun or Tilmun—Sumerian Eden

Migdol, Magdala—Hebrew great pillar stone, tower

Ultima Thule—North polar paradise, the White Island, the land that departs

Tlaghtga—ritual place in Ireland where the sacred new fire was lit

Tollan (Tulan)—place of reeds, Toltec heaven on a mountain as high as the moon

Tuhalha—Mayan sweat bath house

Tlalocan—heaven of Tlaloc, from *tlalli*—earth

Primal Personages

Atalu—Assyrian water serpent (counterpart of the Hebrew Leviathan)

Atlas supported heaven, father of the seven Pleiades and first king of Atlantis. Guards the garden of Hesperides on Mount Atlas.

Daedalus—built the Cretan Labyrinth

Dalai Lama (Tibetan) "Ocean of Wisdom"

Tal—primitive names for Hercules:

Talus (Crete)	Tantalus (Pelasgia)
Telamon (Greece)	Taliesin (Celtic)
Telmen (Syria)	Tailltean (Ireland)

Talos—hurled to death from the first Acropolis

Tlaloc—Veracruz rain god who lived in a mountain garden with four caves

Chapter 11

The Wings of God

As soon as Jesus was baptized, he went up out of the water. At that moment heaven was opened, and he saw the Spirit of God descending like a dove and alighting on him.
——Matthew 3:16

It is evident from our ancestral accounts that the Holy Spirit (Ruach HaKodesh) with wings outstretched is remembered in every part of the world. Although He shines like the sun, He is not the sun but the Spirit of the Living God. In Jewish mystical literature, the Shekhina (Divine Presence) is represented as a bird, a dove among doves, and one of the names for the hidden Palace of the Messiah is the Bird's Nest.[1] His reappearance is expected at the redemption of Israel when everything she has lost will be restored.[2] John Milton wrote of this mystery:

> Thou from the first
> Wast present, and with mighty wings outspread
> Dove-like, satest brooding on the vast Abyss
> And made it pregnant. (Paradise Lost)

The Holy Spirit as a dove within Jubal, thirteenth century
Window of Auxerre Cathedral (John O'Neill 1897)

The Rg Veda calls this holy King of Birds by many different names. Sanskrit scholar Ralph TH Griffith identifies them all with the same divine being, each expressing a different aspect of the One Supreme Spirit:

> They call him Indra, Mitra, Varuna, Agni, and he is the heavenly noble-winged Garutman. To what is One, sages give many a title ... The Bird Celestial, vast with noble pinion, the lovely germ of plants, the germ of waters. (Book 1 Hymn 164 (notes), Ralph TH Griffith, trans. *Hymns of the Rg Veda*

In his *Dictionary of Symbols,* JE Cirlot notes that the giant bird is always symbolic of the Creating Deity. He also brings celestial messages.[3]

In many traditions, the eagle or hawk represents the Supreme Spirit because it hovers "over the face of the waters" (Genesis 1:2). The primitive Chinese character for "hover" is *hsun* ✝. "Its wings do not flutter."[4] To the Greeks, the eagle is one of the oldest symbols of Zeus and is engraved on his altars. Bunjil is the Eagle-hawk Ancestral Father of the Kulin and Wurunjerri groups of Victoria and New South Wales.[5] The Vedic Indra appears in hawk form, as does Apollo as he flies down from Mt. Ida in Homer's *Iliad*. The Roman imperial golden eagle symbolized Jove.[6]

Assyrian Nisrok hovering over the Tree of Life
William Hayes Ward, *The Seal-Cylinders of Western Asia* (1910)

The similarity of names to the Hebrew *ruach* (spirit, breath, wind) is remarkable. These include the Malayan *ruwak-ruwak*, the Persian Shahrokh (King bird), and the eagle Nisrok, the primordial creative deity of Assyria[7] worshipped by Sennacherib (Isaiah 37:38). The Chinese call the divine bird "the Roc."[8] He is called *rukh* or *rokh* in the *Arabian Nights*. The Mayan serpent-eating Uok (hawk) is the messenger of Hurakan, "Maker, Modeler, and Heart of the Sky."[9] The English word rook is both a bird and a chess piece in the ancient symbolic game of good versus evil.

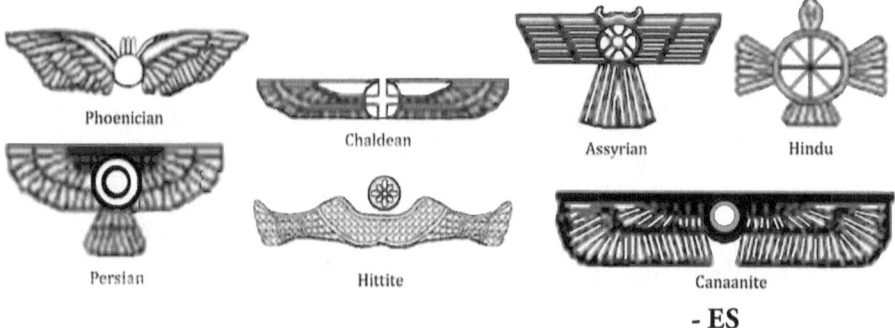

- ES

The Old Norse word *rök* is defined as reason, ground, origin, wonder, sign, and marvel. The All-Father Odin owned a divine hawk skin, and a giant eagle sat at the northern edge of heaven:

> Hraesvelg he is called
> Who at the end of heaven sits,

A Jotan (giant) in eagle's plumage:
From his wings comes
It is said, the wind
That over all men passes.
(Snorri Sturluson, *The Prose Edda* trans. B. Thorpe)

The Phoenix

If I take the wings of the morning, and dwell in the uttermost parts of the
sea; even there shall thy hand lead me, and thy right hand shall hold me.
—Psalm 139: 910 (KJV)

The phoenix symbolized divinely ordained periodic destruction and
resurrection to the Greeks. He built his nest on a pyre of sweet-smelling
wood and burnt himself to ashes, whereupon his son arose from his bones.[10]
The word phoenix means both purple and palm tree. The appearance of
the phoenix was seen as an omen of miraculous events to come. He was
personified in European folktales as Cinderella or Aschenputtel (Ash Fool)
reborn from the ashes of the hearth. Among Native Americans, he is Dirty
Boy or Little Ash Boy.[11] The great phoenix of our time is Israel and the Jewish
people who arose from the ashes of the Holocaust.

The ancient Chinese considered the beautiful Feng-huang (phoenix) 鳳
凰 to be the emperor of all birds. His sweet song was a harmony of five notes.
His wings resembled clouds, and his twelve beautiful tail feathers were a blend
of the five colors.[12] He first appeared at the time of Yú Huang Tiān Shangdi
天玉上帝, a time of peace and prosperity when wonderful and miraculous
events occurred. When he flew away, the country was visited with calamities.[13]

The Persian phoenix had several names, Kamak, who hovered over the
earth with a wingspan so great the rain could not fall.[14] He is also the cosmic
Senmurw (or Simurgh), a green peacock nesting in the Tree of Life on Mt.
Alburz. Here he assists the warrior hero Rustam. The name of the Tree, Hom
(Old Iranian Humaya, Avestan Haoma) is related to the Hebrew word *tehom*
(deep). The tree stands in the middle of the world sea Vourukhasa. When
the bird took flight, the leaves of the tree shook, causing the seeds of every
plant to fall. These floated on the winds of Vayu-Vata and the rains of the star
Tishtrya. Taking root, they cured all the illnesses of humanity.[15] The Russian
version is the firebird, whose glow lit up a tree with golden apples.

In the Kojiki of Japan, Yamato turns into a white bird, for it was commonly believed that the soul resembles a bird. Four songs were sung at Yamato-take's august interment, "So to the present day these songs are sung at the great interment of a Heavenly Sovereign. So the bird flew off from that country, and stopped at Shiki in the land of Kafuchi. So they made an august mausoleum there, and laid Yamato-take to rest." It was called "the August-Mausoleum of the White-Bird." Nevertheless, the bird soared up to heaven again and flew away.[16]

The Benu Bird of Egypt

I long to dwell in your tent forever and take
refuge in the shelter of your wings.
—Psalm 61:4

According to the sacred texts of Heliopolis, the Creator Ra arose in the form of a Benu Bird 𓈖𓏤𓅨𓅱 and hovered over the primeval waters at the dawn of life. The Benu alighted upon the Primeval Hill on the Ben-ben Stone. Then Ra spoke the divine "words of power" that brought all nature into existence.[17] This worldwide connection between the bird and the rock is evidenced in the proto-Indo-European root *petro,* which means bird, wing, feather, and rock.[18]

The word *benu,* from *uben* (to shine), means palm tree, heron, rise, and ascend. The Benu is often pictured as a golden hawk with a heron's head. It was believed to be the soul of Asar and the heart of Ra and was identified with the morning star. Egyptologist Rundle Clark calls the Benu the herald of all things to come and a primeval form of the Creator. The Benu flies from the Isle of Fire beyond the limits of the created world. The place where it landed became the center of the earth, Heliopolis, the biblical On.[19]

All the Creator gods of the various regions of Egypt were represented in bird form, and the Egyptian crown had feather plumes. The hawk glyph is a synonym for Heru (Horus, the Face), the morning star and god of Latopolis, addressed as "he who is above."[20] Ra, Asar, Seker, Temu, and Ptah are all pictured with hawk's heads. In the Book of the Dead, two alternating birds are mentioned. "I go in like the Hawk, and I come forth like the Benu, the Morning Star of Ra."[21] In the same source, Asar speaks using bird imagery,

"I am lord of millions of years. I have made my nest in the uttermost parts of heaven" (Chapter 84). Tehuti (Thoth), Creator god of Hermopolis, is represented with the head of an ibis, and the goose was a very ancient form of the earth god Geb.

Wings over America

Like an eagle that stirs up its nest and hovers over its young, that spreads its wings to catch them and carries them on its pinions. The Lord alone led him … (Deuteronomy 32:11,12a)

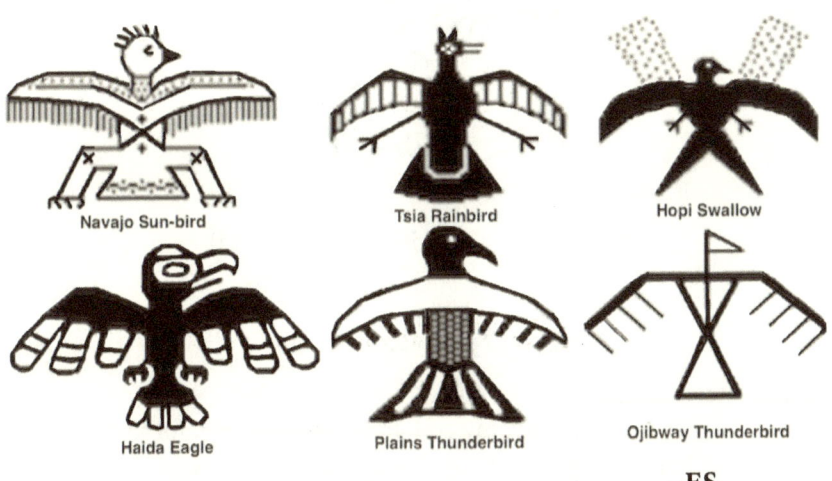

Navajo Sun-bird

Tsia Rainbird

Hopi Swallow

Haida Eagle

Plains Thunderbird

Ojibway Thunderbird

- ES

The Great Spirit of America

The eagle feather is a symbol of power and the breath of life to all Native Americans.[22] The Choctaw venerate the Great Spirit as Heloha, Thunderbird of the Northland. The Cherokee know him as Awahili, the Great Hawk. The Osage call him Henga, the Sacred Eagle. The Abenaki know him as Wind Eagle who perched on a mountain peak. The Sioux call themselves "eagle people," for eagle is the messenger of the Great Mystery. His eyes flashed forth lightning, his wings produced thunder, but he is a benevolent spirit.[23]

The Iroquois call him Oshadagua, the Great Dew Eagle of the Mists. "From his spreading wings falls healing moisture."[24] The Algonquin giant eagle, Thunderbird, prevented the earth and vegetation from drying up and dying.[25] The Quillayute locate Thunderbird's perch in the "Hunting Ground of the Great Spirit."[26] His image crowns the totem pole, which represents the link between the tribe and the Great Spirit. Feather headdresses are worn in imitation of the One, who promised to "cover thee with his feathers ..." (Psalm 91:4).

Mississippian four winds spirit-bird (ES)

Quetzalcoatl

Have mercy on me, my god, have mercy on me, for in you I take refuge. I will take refuge in the shadow of your wings until the disaster has passed. ¯Psalm 57:1

In the beginning, say the Aztecs, there was nothing but water and Quetzalcoatl. He was the most powerful figure in Mexican and Central American religion.[27] He created the first human couple and lived in the highest heaven. Unlike other deities, he did not demand human sacrifice.[28] He was closely associated with the morning star, a preexisting power older than the sun.

His Nahuatl name means both "precious twin" and "feathered serpent." His titles include Ehecatl (the Wind), Ce Acatl (One reed) an epithet of the morning star, and Tolpitzen (Our Revered Prince).[29] Several versions of his death and disappearance exist, and all include his promise to return. His most important symbol is the almost extinct green quetzal bird with its long, beautiful tail. Whole bird headdresses wore worn in his honor.

He was lord of the celestial waters and genius of the night sky.[30] Although he ruled from his great house in the sky, he worked within this world as savior and guide to all humanity. Rulers derived the right to govern from this god who was also believed to be a king.[31] He wore a conical hat and a wind jewel representing the whirlwind. He held an elaborately decorated curved scepter resembling a shepherd's crook.[32] As the inventor of the calendar and the originator of all arts, crafts, and sciences, when he left, culture decayed.[33] The Mayans call him Kulkulkan.

EL Dorado

Because you are my help, I sing in the shadow of your wings.
—Psalm 63:7

The colorful hummingbird is also revered in many places as a symbol of the Great Spirit, especially in the Americas. In Mayan lore, the sun can disguise itself as a hummingbird.[34] The principal god of the Mexica and the Tenochas was Huitzilopochtli (hummingbird wizard of the left). A divine voice issued from a tree that broke in two and fell down. This incident of the broken tree began Aztec history. The voice led them to the promised land of Tenochtitlan (Mexico).[35] The original site of the temple of Huitzilopochtli is now the site of Mexico City's great cathedral.

Ancient tales of a golden city, EL Dorado, lured Spanish explorers again and again to South America. The Muiscan and Chibcha tribes of Colombia revered EL Dorado, a golden hummingbird, and priest-king. EL Dorado was also the name of an invisible heavenly city with streets and palaces of solid gold, whose inhabitants were ideally happy and lived forever. It was said that the city would become visible again at the end of the world.[36] A man (Zipa) covered with gold dust represented the deity in a ritual journey to the sacred mountain lake, Guatavita, accompanied by four principal chiefs. Sailing on a rush raft to the middle of the lake, offerings of gold and emeralds were thrown in.[37]

According to the Mbyá Guarani people of Paraguay, the First Father lived in the primal winds. They said he had a sacred headdress of feathers that were like flowers covered with morning dew. The very first bird, the hummingbird, hovered among the flowers. Before the sun existed, he was lit by the reflection of his own inner self. The thoughts inside of his sacred being were his sun. His paradise and a new space and time will be resurrected in the future.[38]

The Great Spirit Recalled as a Raven

> May the LORD repay you for what you have done. May
> you be richly rewarded by the LORD, the god of Israel,
> under whose wings you have come to take refuge.
> —Ruth 2:12

Tulungersaq (Father Raven) is the Creator of all life and the Great Spirit worshipped by the Inuit of Alaska and tribes of the northwest coast of North America. He made the first people out of clay and said to man one day, "You are lonely. I will make for you a companion." He took some white clay and made a figure like a man.[37] Father Raven was no ordinary bird. He is the origin of everything, and his spirit power is holy. When he finished creating the earth, Raven called all human beings together and said, "I am your Father and to me you owe the land you have and your being, and you must never forget me."[38]

The Inuit say Father Raven became angry at the people for killing everything he had made:

> Raven went away and took the sun out of the sky. He put
> it in a skin bag and carried it far away, to a distant part of
> the sky-land. Then it became dark on earth. The people on
> earth were frightened when the sun vanished. They offered
> Raven presents of food and furs if he would bring back the
> sun. Raven said, 'no' but after awhile, Raven felt sorry for
> them, so he let them have a little light.
> (Katherine Berry Judson, Ed. *Myths and Legends of Alaska*)

According to the Tlingit, Haida, and other tribes, Raven was once pure white. The Athapascan of Upper Yukon say that Raven once had many colors but later turned black.[39]

The name of the British tribal god Bran (or Vron) means "crow" or "raven." His body was as big as a mountain. He was called Bran the Blessed, as he was able to restore life to the dead. At the final battle, he was wounded in the foot by an arrow, and he ordered that his head be cut off. White Hill, the site of the Tower of London, is the burial ground of Bran's singing head. It was believed to protect the city from invasion.[40] As far back as anyone can remember, there have been tame ravens at the Tower of London with an official protector. It is said that the security of the crown depends on the birds' continued presence. The crow was also the messenger of the Celtic god Lugh, once widely venerated in Ireland.

The raven was the oracular bird of Odin. Dried raven hearts were crushed and applied to cubic dice to bring good luck.[41] The Crow is also the cultural hero and ancestor of the Kulin nation in central Victoria (Australia), where he is known as Waa, Wahn or Bellin-Bellin. He was in charge of the winds and caused a whirlwind so strong it blew the Eagle-hawk ancestor Bunjil and his people into the sky world.[42] It is obvious in the words themselves that the raven, crow, dove, owl, sparrow, swallow, hawk, and eagle (Latin *aquila*, Hebrew *okhel)* were sacred birds.

Taking the Auspices

How priceless is your unfailing love, O God! People take refuge in the
shadow of your wings.
—Psalm 36:7

Aside from land surveying, the duties of the Roman augur included taking the *auspices* (divination using the flight patterns of birds). No private or public transaction ever took place without this rite. The word *auspex* derives from *avis* (bird) and *spec* (look). In Rome, the eagle and the vulture were regarded as *Jovis ales* (Jupiter's birds) and his messengers.[43] Romulus was said to have been the first and best augur. From him all succeeding augurs received the *lituus,* his curved staff.[44]

Holding the lituus in his right hand, the Augur found a center (Templum) and drew a cross in the sky. In each quarter, he looked for birds. Any sign in the northern half was significant, the northeast being the most favorable; the southwest (down among the dead men) was the most unfavorable. Only certain birds were auspicious, the Oscines by singing and voice, and the Alites by flight. The birds used were the raven, crow, vulture, owl, and the green woodpecker of Mars. By the time of Cicero (106 BC–43 BC), who served as state augur, taking the auspices had degenerated into watching caged hens eating.[45]

Lituus on Roman denarius Crozier of archbishop (ES)

The belief that birds revealed the will of the gods was common to many nations. We still use the words auspicious and inauspicious today. The Greeks took auspices, but eventually the oracles supplanted the birds, and the future was learned from Apollo and others, rarely from Zeus, who possessed few oracles.[46] Lituus is an Etruscan word that derives from the Semitic *littu,* which in both Babylonian and Hebrew (*lahat*) means "flaming." It is used in describing the action of the Ever-turning Sword in Genesis 3:24. Augury, forbidden in Israel, strangely survived in the Church. The *littus* is the prototype of the bishop's crozier. The augur's sign of the cross in the air is identical to the sacerdotal blessing.[47]

The Bird of Storms

> But to you who fear my name the Sun of Righteousness
> shall arise with healing in His wings.
> —Malachi 4:2a (NKJV)

A dark bird, a counterfeit of the Holy Spirit, is blamed for the Flood by many tribes. These include the Hebrew Ziz, whose wingspan eclipsed the sun, the Sumerian Imdugud, a lion-headed eagle, and the Akkadian Anzú (often shortened to Zu). The name Anzú means "heaven" and "know." He was the servant and doorkeeper of Ellil with a beak like a saw. His name, Zu, was part of the cuneiform script used to write the word mist.[48] He provided the earth with moisture but then disrupted the kingship of heaven and tore the sky apart with his talons.[49] He assaulted the kingdom of Enlil and stole the Tablets of Destiny, bringing on total darkness,

> He stole the Ellil-power; rites were abandoned.
> Father Ellil, their counselor, was dumbstruck
> Radiance faded (?), silence reigned,
> Every one of the Igigi (gods) was thrown into confusion,
> For he had stripped the chamber of its radiance.
> (Anzú II Old Babylonian Version)

Anzú's nest was in the sacred Halub tree, and his wings were said to cause whirlwinds.[50] The speech of the Zu bird resembles Lucifer's in Isaiah 14:11–14 with all his "I wills." We find it in the Babylonian Second Tablet of Creation:

> His eyes saw the mark of rulership,
> The crown of his sovereignty, the garment of his divinity.
> Zu saw the divine tablets of fate.
> He looked at the father of the gods, the god of Dur-an-ki,
> Desire for rulership seizes hold of his heart.
> 'I will take the tablets of the gods
> And decree the decisions [of all the gods.]
> I will establish my throne, I will proclaim laws.
> I will give all orders to all the Igigi.'
> Zu proceeds to the dwelling-place of Enlil and waits for a
> favorable moment to make an attack.
> His heart was bent on the contest.
> With his gaze directed toward the entrance of the dwelling,
> He waits for the beginning of day.
> As Enlil poured forth the brilliant waters,
> Took his seat on his throne and put on his crown,
> He snatched the tablets of fate out of his hands,
> Seized the authority—the promulgation of laws."
> Thereupon Zu flew off and hid himself in his mountain."
> (Trans. by Morris Jastrow)

In India, it was Garuda, chief of the feathered race, who ended the Golden Age. They praised him as the highest being and called him the fire and the sun. The name also means emerald. The Mahabharata describes the time when he brought the great dust storm that altered the sky and darkened the worlds. He brought the heavens into such total disarray that he could no longer be seen. He confounded the gods and guardians under the tent with his talons and beak, and with the wind of his wings he tore the heavens asunder.[51]

The Mayans blamed the killing of a cosmic bird for the Flood. The giant parrot Vucub-Caquix (Seven-Macaw) once dominated the heavens as the sun and the moon. "His light provided a sign for the people who were flooded."[52] With his red feathers, shining white nose, silver and emerald eyes, and perfect teeth, he was "like the face of heaven." He lived high up in a fruit tree called "the Great Tree of Seven Macaw." The scope of his face lived right around his own perch. As "the foothold and walkway of the people," he was "the pivot of the movement of the night sky" and "the celestial light at the Pole-star."[53]

Knowing his strength lay in his teeth, Seven-Macaw's enemies, the twins Hunahpu and Xbalanque, shot him down from his perch with a blowgun. They pulled out his beautiful blue teeth resembling jewels, tore off his lower jaw, and gouged out his shiny metal eyes until all his beauty was gone. He no longer looked like a lord. At his fall, mountains were raised and leveled in a single day. Then came the great rain that destroyed the people for whom he had been the sun. The twins then became the sun and moon of the present creation. The remains of Seven-Macaw were said to be the seven stars of Ursa Major. His image appears on a northern tree at Palenque in the central panel of the Tablet of the Cross.[54]

Aztec lore refers to the time when the sky crashed down upon the earth and the Lord of the Night stole the sun, plunging the earth into darkness that lasted for a long time. The following prayer by an Aztec priest is quite moving:

> O mighty Lord, under whose wings we seek protection, defense, and shelter! Thou art invisible, impalpable, as the air and as the night. I come in humility and in littleness, daring to appear before thy majesty. I come uttering my words like one choking and stammering; my speech is wandering, like as the way of one strayeth from the path and stumbleth. I am possessed of the fear of exciting thy wrath against me rather than the hope of meriting thy grace. (Prayer of an Aztec priest, HB Alexander, *Latin American Mythology*)

In several American accounts, the devil is pictured as the ferocious ruler of bats, Camazotz, with teeth and claws and a nose like a flint knife. He beheaded one of the hero twins as they were contending against the powers of evil.[55] After his assault, this usurper assumes the throne of heaven for a brief interval of seven years according to some accounts. The Egyptians called this "the time of troubles" (see Ecclesiastes 3:15 AMP).[56]

The Choctaw tell of a great buzzard with a glorious topknot. All the feathers on his head were lost because of his pride.[57] The Chane of Guiana say that in former times, fish swam in the branches of a huge yuachan tree, but the Trickster shot down a large dorado in the tree, flooding the entire world.[58] According to the Pehuenche of Tierra del Fuego, the Flood was a turning point in cosmic history. It was accompanied by a long period of darkness when the sun and moon fell from the sky. This lasted until two giant condors brought the present sun and moon into the sky.[59]

A similar account from Kimberly, Australia, tells of a group of children who tortured Dumbi, the great sacred owl during the Dreamtime. He managed to fly away but complained to the Wandjina (the primal ancestors), who became angry. The waters rose and drowned all the people except two, a girl and a boy, who became the ancestors of the present human race.[60] Like the eagle of the Edda that sits in the crown of the World Tree, the tribes of Central Asia (Yakuts, Buriats, etc.) describe a giant eagle, Garide, pecking at the tree of life. When flying furiously, he caused storms, thunder, pitch-black darkness, and a mighty whirlwind.[61]

One question we have not yet answered. How did our earliest ancestors recall seeing the Holy Spirit hovering over the face of the waters? These accounts are much older than Moses. Only two people actually saw this wondrous event—Adam and Eve—right after their exile from God's Earth to our present earth. After their exile, they witnessed the recreation of the sky, the establishment of God's Throne on high, and His triumph over chaos.

According to the Dakota Sioux, the Father promised to return:

The Father will descend,
The earth will tremble
Everybody will arise,
Stretch out your hands.
The Crow-*Ehe'eye!*
I saw him when he flew down,

To the earth, to the earth
He has renewed our life …
(Margot Astrov, *American Indian Prose and Poetry*)

Native American Names for the Great Spirit

Algonquin	Kitche Manitou
Athapascan	Qawaneca
Apache	Naiyenesgani
Aleut	Agugux
Choctaw	Nanih Waiya
Cheyenne	Heammawhio
Chinook	Ikanam
Cherokee	Asgaya Gigagei
Chickasaw	Ababinili
Fox	Ketchimanetowa
Hopi	Taowa
Iroquois	Ha Wen Neyu
Mayan	Kulkulkan
Aztec	Quetzalcoatl
Lakota	Wakan Tanka
Omaha	Wakanda
Oglala	Nagi Tanka
Pawnee	Tirawa Atius
Powhatan	Okeus
Tlingit	Nascakiyel
Ute	Sunawavi
Winnebago	Wahhahnah
Yakima	Wheemeemeowah
Yamana	Watawinewa
Zuni	Awonawilona

Chapter 12

The Rock of God

The Mystery of the Chief Cornerstone

> He is the rock, his works are perfect, and all his ways are just. A
> faithful God, who does no wrong, upright and just is he.
> —Deuteronomy 32:4

The Lord is called "the Rock" many times in scripture. To this Rock are anchored the foundations of the earth. The Lord asked Job about this, "Where were you when I laid the earth's foundation? Tell me, if you understand. Who marked off its dimensions? Surely you know! Who stretched a measuring line across it? On what were its footings set, or who laid its cornerstone while the morning stars sang together and all the angels shouted for joy?" (Job 38:4–6).

All early tribes remembered the Foundation Stone visibly shining from the center of the Throne of God and connect its removal with flood and disaster. Thomas Burnett recognized this fact in 1691:

> A foundation, which will bear the weight of two worlds
> without sinking, must surely stand upon a firm rock.
> (Thomas Burnet, *Sacred Theory of the Earth*)

Vail also pointed this out in 1874:

> We recall that the ancient earth had 'four corners,' and it
> is not difficult to find the original Chief Cornerstone in

the luminous north. (Isaac N. Vail, *The Waters above the Firmament*)

The English word *rock* derives from the universal root *arak* in the holy tongue meaning "pole," from which also derives the Hebrew word *rakia* (firmament). A heavenly rock seat or stone throne is found in all ancient traditions. In Israel, the Even haShetiyah is the Foundation Stone. According to the Talmud and Midrash, the Holy One created the world from Zion after casting this stone into the sea.[1] The Lord "fixed it" over the mouth of the tehôm, the great deep, and engraved upon it His own ineffable name. When the flood generation sinned, God removed His chief cornerstone, and the entire canopy collapsed, flooding the earth.[2]

In both biblical and tribal symbolism, the Creator is identified with a sacred rock that is all of the following:

A foundation stone	A Son-stone
A living stone	A sun-stone
A pillar stone	A navel-stone
A healing stone	A prophetic stone
A most precious stone	A fixed stone
A pivot stone	An exiled stone
A magnetic stone	A swallowed stone
A cubic stone	A stone that cries out
A witness stone	A fertility stone
A rain-stone	A birthstone
A stone of destiny	A stone of help

The Tribal Foundation Stone

And all the men of Shechem gathered together, all of Beth Millo, and they went and made Abimelech king beside the terebinth tree at the pillar that was in Shechem.
—Judges 9:6 (NKJV)

When the Foundation Stone disappeared at the Flood, all tribes looked for a replica on earth to place at the navel of their land. This stone became the center of tribal life, the place of assembly, and the coronation site of kings. God gave Abraham the promise of the land at Shechem, and a pillar stone was set up. The Mizpah stone memorializing the covenant between Jacob and Laban became the navel center of the twelve tribes. Later, "Samuel took a stone and set it up between Mizpah and Shen, and called its name Ebenezer (Stone of Help), saying, 'Thus far the Lord has helped us'" (1 Samuel 7:12 NKJV).

The choice of the tribal stone was of utmost importance. A huge, unhewn green stone, such as jade or emerald, was ideal because God created it, and it was the color of His city. Inca historian Garcilaso de la Vega (1539–1616) wrote:

> At the time of the Spanish Conquest, an immense emerald, almost as large as an ostrich egg, was adored by the Peruvians in the city of Manta. This emerald goddess, bore the name of Umina, and like some of the precious relics of the Christian world, was only exhibited on high feast days, when the Indians flocked to the shrine from far and near, bringing gifts to the godess. The wily priests especially recommended the donation of emeralds, saying that these were the daughters of the goddess, who would be well pleased to see her offspring. In this way an immense store of emeralds rewarded the efforts of the priests.
> (*Comentarios Reales de los Incas*)

A fallen stone, such as a meteorite or a magnetic stone like loadstone, was also ideal. One came from heaven, and the other pointed there. The word *calamity* derives from the Hebrew word *kalamitah* (magnet) found in Proverbs 21:23, "Those who guard their mouths and their tongues keep themselves from calamity." *Calamita* is still the word for magnet in Italian. Olaus Magnus mentions a magnetic northern mountain, Monte Calamitico, surrounded by many wrecked ships. This mountain drew all iron nails and weapons to it. Ships were constructed with wooden pegs because of this archaic belief.[3]

Four revered meteorites were the Palladium that fell from heaven at Troy. The city's safety depended on the stone's remaining there, but Odysseus and Diomedes carried it off. Others were the Stone of Ephesus and the Black Stone of Pessinus taken to Rome during the last Punic War.[4] It is said that the Mongolian black oracular stone sent by the King of the World to the Dalai Lama was of this type. It was once believed that if the tribal stone was removed from its original site, water would gush out, submerging the earth, a tradition still attached to the Foundation Stone of the temple in Jerusalem.[5]

Idolatry of Stones

And do not erect a sacred stone, for these the Lord your God hates.
—Deuteronomy 16:22

The Gilgals of Jacob, Moses, and Joshua with their twelve pillar stones later became places of apostasy. "Seek me and live; do not seek Bethel, do not go to Gilgal, do not journey to Beersheba. For Gilgal will surely go into exile, and Bethel will be reduced to nothing" (Amos 5:4b–5:5; see also Amos 4:4, Judges, 3:19, 1 Samuel 7:16). Under King Jeroboam, Bethel became a rival of Jerusalem. "Jeroboam ordained a feast on the fifteenth day of the eighth month, like the feast that was in Judah, and offered sacrifices on the altar. So he did at Bethel, sacrificing to the calves that he had made. And at Bethel he installed the priests of the high places which he had made" (1 Kings 12:32).

As the generations passed, traditions of the Foundation Stone above and its tribal replica became indistinguishable. At first, natural unhewn stones were used in rituals, but eventually carved stones were set up, shifting the focus from the person of God to the stone itself. The central upright menhir, called the sunstone, was believed to be inhabited by the ancestral god.[6] The Egyptians fashioned their pillar stones into obelisks (Jeremiah 43:13). Prophets denounced them, and faithful kings removed them. "(Hezekiah) removed the high places, smashed the sacred stones …" (2 Kings 18:4; see 2 Chronicles 14:3, 4). An enemy capturing or breaking up the tribal stone would totally demoralize a people.

The tribal stone was widely used in fertility rites. Women desiring children slept on it, rubbed against it, or slid down it. As recently as 1880 in France, during the full moon, a naked husband chased his naked wife around an ancestral stone in order to conceive.[7] The dead were buried around the stone, oaths were sworn on it, and the sick laid near it. Anointed with oil and smeared with the blood of sacrifices, the people wore down its surface with kisses (the Irish Blarney Stone probably began this way). Here, trials were held to determine the fate of criminals.

Most tribal gods originally began as an uncarved stone. The Greek geographer Pausanias (ca. AD 110–180) wrote:

> At a more remote period all the Greeks alike worshipped uncarved stones instead of images of the gods (argoi lithoi). (*Description of Greece* trans. WHS. Jones)

Mithra, an ancient deity of both Persia and India (Mitra), was born of a stone on December 25, and one was honored in his cult.[8]

A stone represented Herakles (Hercules) in his temple at Boeotia, and an unhewn stone was the most ancient image of Eros.[9] Apollo, whose epithet was Lithesios (from *lithos* stone), appears to have been represented by the most stones.[10] One was shaped like a small pyramid, and another was in the form of a pillar. His pointed pillar stones called Agyieus were set up by courtyard doors.[11]

Hermes, the Etruscan god of crossroads, the pivot, and the doorway, who claimed to have invented the bow drill and the alphabet, was originally just a stone monolith. The books of Hermes were said to be "the divine fountain of all true knowledge of the Mysteries."[12] Pillar stones (*herms*) were set up at boundaries to promote fertility.[13] When Hermes ceased to be one with the stone, his appearance became human, and his theophany became a myth.[14]

Ganesh, the popular elephant-headed Hindu god of crossroads, was worshipped as a red stone.[15] The Baltic Slavs worshipped Svantovit as god of gods, father of the sun, fire, and plenty. His image was a pillar with four faces, one at each of the four directions. His attributes were an enormous sword, a bull's horn filled with wine, and a white horse. Rites included the sacrifice of horses.[16] Mercury was originally nothing more than a stone cube. Macrobius wrote that Mercury was represented as a square block with a head and phallus.[17]

Oracular Stones

> They say to wood, "You are my father," and to stone, "You gave me birth." They have turned their backs to me and not their faces; yet when they are in trouble, they say, "Come and save us!"
> —Jeremiah 2:27

Above all, the tribal stone represented communication with God. A prophet *(navi)* stood on the navel stone to proclaim the Word of God. The Hebrew word *nava* means "to prophesy, bubble forth, pour out, utter." The Vedic word Pranava is "the Creative Word of God."[18] These navel oracles were once so plentiful that the Greek poet Pindar wrote of "the templed omphalos of oracle-roaring earth."[19] Wars broke out over their possession, and whole nations were led astray by their ambiguous answers. Herodotus wrote that the Pelasgians only consented to the Olympian system at the command of the Dodona Oracle.[20] The Corinthians received an oracular order to worship a pine tree as a god.[21]

The most famous navel stone of ancient times was the Omphalos at Delphi directly below Mount Parnassus. It was believed to be the center of the world. Aeschylus wrote:

> It remains even into the third generation, ever since Laius—in defiance of Apollo who, at his Pythian oracle at the earth's center, said three times that the king would save his city if he died without offspring ... (*Seven Against Thebes* trans. HW Smyth)

The Greeks swaddled the stone with woolen bands, anointed it daily with oil, and called it Abadir (Glorious Father). Apollo was its most recent possessor, but it originally belonged to Gaea (earth).[22]

The Delphic Oracle was allegedly the stone Kronos swallowed. According to Greek myth, he swallowed his children to prevent the birth of a greater king than him. To save Zeus, Hestia swaddled a stone like a baby and gave it to him. He ate the stone, and it *stuck in his mouth*. When he was grown, Zeus gave his father a potion, causing him to vomit up the stone and the rest of his children, then drove him from the sky.[23] A crude explanation of the changing heavens at the Flood, it arose from the memory of the chief cornerstone "fixed" in the "mouth" of the rivers. Demons, speaking through idols and oracles, twisted the tale into the lie that God desired child sacrifice, a practice vehemently condemned by the Lord through all His prophets (see Ezekiel 16:20).

The swallowed stone

By Jupiter, the Stone

Is there any god beside me? No, there is no other
rock; I know not one.—Isaiah 44:8

In the Roman version of the tale, Jupiter was the stone swallowed by his father, Saturn. According to French comparative mythologist Georges Dumezil, the name Diou or Iou existed among all Italic peoples and has the meaning of "shining heaven."[24] Like the Vedic Dyauh Pita, Jupiter means "god, the father," and Lapis (stone) is one of his titles. The most binding and sacred Roman oath was Per Iovens Lapidem (By Jupiter the Stone). According to Festus, the tradition required those who swore by Jove to hold a flint axe (Lapis silex).[25] This was carried through the city in times of drought, so that the Master of the Rain would accommodate.[26]

Since the Foundation Stone covered the deep, it was seen as the "cover" of the abode of the ancestral dead. When a new camp was established, the Romans dug a round hole, and a stone called Lapis Manalis was placed at the bottom. This was seen as the gate of the underworld. Several times a year, the stone was removed for the *manes* (ancestral spirits) to pass through. According to Plutarch *(Roman Questions),* blood sacrifices were offered to the *manes* or Di Parentes several times a year to appease their anger.[27] The feast of the Parentalia was celebrated from February 13 to 21. On the last day, the Feralia, temples were closed, fires did not burn, and marriages were not contracted.[28]

Egypt, the Benben

There in the temple of the sun in Egypt (in Heliopolis),
he will demolish the sacred pillars ...
—Jeremiah 43:13a

The Egyptian Foundation Stone was the Benben. The temple of the Benben was the center of calendrical rites, where the mysteries of creation were read and the damaged parts of the divine eye mystically restored. Pilgrimages were made to a replica of the Benben, a pyramid-shaped obelisk in a temple called the Het-Benben (House of the Foundation Stone).[29] The capstone of the pyramid was its representation.

The Pyramid Texts indicate how sacred it was to the people:

> O Atum! When you came into being you rose up as a High Hill. You shone as the Benben Stone in the Temple of the Benu in Heliopolis. (Utterance 600 trans. Kurt Heinrich Sethe)

According to the texts, at the dawn of life, Ra as a Benu bird alighted upon the Benben Stone. From there, he spoke the divine "words of power" that brought all nature into existence.[30] The gilded surfaces of the Benben reflected the morning sun.[31] The word is related to *banban* (to flood), the Aramaic *eben* (stone), the Babylonian *banu* (to build, create, to shine), and *labanu* (to make bricks), the meaning of the name Laban.

Mesopotamia, the Elmesu Stone

He lifted me out of the slimy pit out of the mud and mire; he
set my feet on a rock and gave me a firm place to stand.
—Psalm 40:2

The Sumerian Foundation Stone was the Elmesu Stone. Described as precious, wonderfully wrought, and perfect in celestial beauty, it disappeared at the time of the Flood. Its likeness was worn as a shining breastplate on the chest of the king.[32] In the texts, its loss, along with the Mesu tree, is said to have caused absolute devastation:

I changed the location of the Mesu-tree (and) the Elmesu-stone, and did not reveal it to anyone. (Erra and Ishum 1, *Myths of Mesopotamia* trans. Stephanie Dalley)

The Babylonians called this same sacred stone Mashu ✝.[33] The most important related Hebrew word is *mashah* (to anoint), from which Moshiach (Messiah) derives. Elmesu, Mashu, and *mashah* are also related to the Egyptian word *mes* 𓄟 (to bear, give birth to, child, son). The first glyph 𓄟 is the original, the second 𓄟, is what it changed into over thousands of years, showing that the scribes no longer understood its significance. Mes is the root of the name Moses and the names of pharaohs, Rameses, Thutmoses, Ahmoses.

The Sacred Stones of the Celts

For the stone will cry out from the wall and the
beam from the chamber will answer it.
—Habakkuk 2:11

In early tribal days, the people gathered at Uisneach, "the Navel of Ireland," where the Druids held court well into Christian times. In the center was the famed Liag Fail or Stone of Fál (Destiny). Ireland calls itself Inis Fail (Island of Fál) after this stone. The Archdruid spoke the *logh* (divine word) when he stood on it. Here the sacred fire was lit by wheel, spindle, and tow.[34] On the Feast of Tara, the king was chosen and inaugurated. When the rightful king stood on this stone, its shriek was said to have been heard throughout the land.[35]

Crom-cruaghair[8] was a rectangular stone once worshipped as chief of all gods. An inscription in Ogham describes the stone as covered with gold and silver. Encircling it were twelve smaller stones covered with brass. In the *Annals of the Four Masters,* Dinnsenchus, the geographer, asserted that this was the principal idol of all the Irish colonies, from the earliest period to the time of St. Patrick. They offered it the firstlings of their children, animals, and corn. St. Patrick struck the image with his Staff of Jesus, calling aloud

---

[8] In Welsh, Pen Crug (Chief of the Mound), Old Persian, Kerum Kerugher.

for the devil to come forth from it. The idol fell to the west, and the twelve surrounding stones sank into the ground.[36]

The Celts of Cornwall revered the Logan Stone, its weight estimated at twelve tons.[37] People with the surname Logan once had ancestors living near it or descended from a prophet. In Madern, Cornwall, people with back and limb ailments and women wanting to conceive crawled through a stone called Mên-an-tol (hole stone), and newborn babes were passed through it in a ceremony of rebirth. All Celts used these granite stone rings—hag-stones (holy stones)—as panaceas.[38] On the continent, in Amancy, La Roche (the Rock) region, France, there is a Middle of the World Stone. As with other tribal stones, it was said to be the only one not covered by the Flood.[39]

Mên-an-tol, Cornwall, United Kingdom
Drawing by John Thomas Blight (1835-1911)

The Stone of Destiny

He opened the rock, and water gushed out;
like a river it flowed in the desert.
—Psalm 105:41

Some Foundation Stones are still used in court rituals, but their history is disputed or no longer understood. When Queen Elizabeth II was crowned in 1953, like all monarchs before her, she sat on a throne constructed over a twenty-six-inch rock called the Stone of Destiny. Some said it was the Rock of the Ancient of Days that fell from heaven at the Flood. Others claimed it was Jacob's pillow taken to Ireland by way of Spain, where it became the Stone of Fál. Captured by the Scots in the fifth century, it was moved to the Abbey of Scone, where it became the coronation stone of the medieval Scottish kings. Edward I brought the stone to England, where it was installed in Westminster Abbey.[40] After years of requests, it was returned to Scotland in 1998.

Throne of Edward I of England 1885.
The Stone of Destiny
fit into the gap beneath the seat.
From *A History of England* (Anonymous)

The British Isles have many ancient standing stones beyond those in stone circles. In *The Old Straight Track,* author Alfred Watkins lists some of these with colorful and interesting names like Bellstone, Hell Stone, Hele Stone, Grimstone, Crick Stone, Tingle Stone, Whittle Stone, Hurl Stone, Pecket Stone, Copstone, Ludstone, Rudstone, Bambury Stone, Kenward Stone, Hangman's Stone, Huxster Stone, Pedlar's Stone, Golden Stone, Blackstone, Whitestone, and several King Stones. The word *the* is used with all of them.[41]

That the British once believed their land to be God's footstool is evident in the eloquent words of Shakespeare spoken by John of Gaunt.

> This royal throne of kings, this scepter'd isle,
> This earth of Majesty, this seat of Mars,

This other Eden, demi-paradise;
This fortress built by Nature for herself,
Against infection and the hand of war,
This happy breed of men, this little world,
This precious stone set in the silver sea,
Or as a moat defensive to a house,
Which serves it in the office of a wall,
Against the envy of less happier lands;
This blessed plot, this earth, this realm, this England,
This nurse, this teeming womb of royal kings,
Fear'd by their breed, and famous by their birth.
(*Richard II*, act 2, scene 1)

The Swedes revered millstones called *alv-kvarnar*.[42] The turning millstone with its square central hole was highly symbolic of the Kingdom of God to all ancient peoples. The new king of Sweden was crowned at Uppsala on a round stone called the Mora stone. Twelve lesser stones surrounded it reflecting the position of Odin, his eight sons and four companions. This was Odin's Rock of Joy.[43] For an interesting look at a sacred millstone in a temple in Egypt.

The Sacred Stones of India

He also brought streams out of the rock, and
caused waters to run down like rivers.
—Psalm 78:16 (NKJV)

Hindus believe stones possess *shakti* (life power) and add force to a prayer, curse, or oath.[45] Stones, *pitr*, are buried in fields, gardens, and houses as guardians at the center and four directions. A stone on an isolated hill is believed to avert the evil eye.[46] Stones add strength to marriage vows. After the bridal pair throws an offering of ghee (clarified butter) into the fire, they circle it, and the bride stands on a stone. This is repeated three times.[47] Stones called *usargallu* (breathing stones) are said to have power to cure sprains and snakebite. Heaps of stones *(kuriya)* are placed on hills and at crossroads.[48]

A stone known as the Stone of Life—*ásma* or "spirit stone"—must never come in contact with the ground. The word *ásman* also means "heaven" in Sanskrit.[49] Brahmins perform a rite called Ásmarohan (standing on a stone).[50]

The worshippers of Shiva, the ancient Vedic storm god of both generation and destruction, regularly pour milk over his pillar-stone idols called *lingams*. They are venerated in temples throughout India as symbols of his creative power and energy.

Sacred Rocks in America

(The Lord) will say: "Where are their gods, the rock they
took refuge in, the gods who ate the fat of their sacrifice …
Let them rise up to help you! Let them give you shelter!
—Deuteronomy 32:37, 38

In Mesoamerican lore, stones fell from heaven, among them a block, which became the altar of sacrifice. Yucatec Mayan books containing traditional knowledge, some preconquest, speak of stones of grace:

> When there was neither heaven nor earth. Then God the
> Father spoke alone, by his own efforts, in the darkness that
> clung like a thrice-withered fruit. This was the first word
> of God, when there was neither heaven nor earth, when he
> came out of the stone and fell into the second stone. Then
> it was that he declared his divinity. Then resounded eight
> thousand katuns at the word of the first stone of grace, the
> first ornamented seven stone of grace.
> (The Chilam Balam of Chumayel trans. Ralph L. Roys)

A swallowed stone also appears in origin accounts. While sweeping, the wife of Mixcoatl swallows a *chalchihuitl* stone (emerald) and becomes pregnant. She gives birth to Quetzalcoatl, whose appearance blocks out the stars.[51] Quetzalcoatl and Tezcatlipoca played a ballgame to determine who would be the sun. The north-south oriented ball-court Tlachco symbolized the night sky and underworld. Divided into four equal, differently colored sections, the central round stone, Itzompan, represented the midpoint of the sky. The captain of the losing team was beheaded over it. Before important games, a priest rededicated the court at midnight. The ball was *uol*, from Nahuatl *ulli* (rubber) or *olin* (motion).[52]

Before the Incas came, the natives of Collao Peru described the maker of all things and father of the sun. He "caused streams to flow from the living stone" and spoke to them with great loving kindness, healed their diseases, and gave them many other benefits.[53] They worshipped him as Inti, the sun star who governs all the stars of the firmament. The conquistadores destroyed his pillar stones set up on the summit of their fortresses. One of these was Tiahuanaco, originally named Paypicala, which in the Aymara tongue means "the stone in the middle."[54] A surviving stone called Intihuatana (hitching post of Inti) in Machu Picchu was tethered to Inti.[55]

The annals of the Cakchiquels of Guatemala contain the traditions of their ancestors. They say the Creator who dwells in the zenith made and sustained man from the obsidian stone in Tulan. They call its replica Chay Abah, which sounds like "Father of Life" in Hebrew. This stone was the oracle of the nation, revealing the will of the gods on important matters, such as when to fight wars and indicating guilt or innocence. The priests and chiefs alone were allowed to gaze on its polished surface. The stone came from an unseen place called Xibalbay (invisible place).[56]

Sacred Stones of the Far East

For their rock is not like our rock, as even our enemies concede.
—Deuteronomy 32:31

The Kashima Shrine near the Kita-ura Lagoon has a Shinto temple founded in the age of the gods. Here stands a worn pillar whose foundation is believed to be the center of the world. Known as the Pivot Stone, it was said to have a restraining influence on the Japanese Leviathan Namazu. Its wild thrashings were said to cause earthquakes and tsunamis.[57] To prevent them, the God-Superior ordered the Daimyojin to hammer the Pivot Stone through its head. Called the Rivet-rock of the World (Kaua-mi-ishi), it is shaped like the navel stones of the West.[58]

Sacred rocks are called *iwakage,* and rock seats are *iwakura.* These are covered with salt by the sick and the devout.[59] The Japanese word for stone is *shaku,* and *ji-shaku seki* is loadstone. Is it possible that the word *shaku* is related to the English *shock,* the Hindu *shakti* (divine power), and the Egyptian *shekem* (divine power)? The Kojiki describes the Heaven-Shining-Great-August-Deity closing and locking the door of the heavenly Rock-Dwelling and retiring:

Then the whole Plain of High Heaven was obscured and all the Central Land of Reed-Plains darkened. Owing to this, eternal night prevailed. (The Door of the Heavenly Rock-Dwelling, part 3, trans. BH Chamberlain)

The Kayah of central Borneo describe their primal deity as the spider-lord into whose web fell a stone. This stone became the earth in which a tree took root, from whose branches grew the ancestors of men.[60] There is universal agreement that Qat, the primal hero of New Hebrides mentioned in chapter 10, was born on the island of Vanua Lava when his mother, a stone, burst asunder.[61]

In the Society Islands, it is said that the heaven god Taatoa embraced a rock, which was the foundation of all things, and so produced the earth and the sea.[62] In Tibetan tradition, a stone from a distant star belonging to Shambhala is called the "Stone of Exile." The greater portion of it remains there, while part of it circulates through the earth, retaining its magnetic link with the main stone.[63]

The Semang people of the Malay Peninsula speak of an enormous stone called the Batu-Ribn that stands at the center of the earth. This rock led to the door of heaven. They say that communication with heaven was once natural and easy, but because of "a ritual fault," only shamans and priests can communicate with heaven now. And this can be done only temporarily.[64]

The Ka'aba

Unless the Lord build the house, they labor in vain that build it.
—Psalm 127:1a

Muslims use Polaris to indicate the direction of Mecca to which they offer prayers. In the center of Mecca is a four-sided black cube, the Ka'aba, which houses a meteorite. It is set on a silver plate and covered with the black pall, the *tob-el-kaaba* "the shirt of the kaaba."[65] Every male must make a pilgrimage (Hajj) to the Ka'aba at least once in his life to circumambulate it seven times. This is called the *tawaf* (circuit). On the day of sacrifice, seven stones are cast onto a cairn, and on the three days following it. The Ka'aba is the oldest, most venerated Arab sanctuary, predating Mohammed by thousands of years.

According to Herodotus, the Arabs once had the custom of smearing their blood on a sacred stone to make a covenant.[66]

The Hajj at Mecca (1907)

The Zoroastrians also have a Kabah, Naqsh-I Rustam, built by Darius in 500 BC. The size of a three-story building, it has only one chamber, reached by one external staircase of thirty steps.[67] The Nabataeans were a North Arabic people whose ancient capital was Petra. They revered a cubic stone with a gold base called *ka'bu*.[68]

> According to the traditions of the Malay Peninsula,
> God created the pillar of the Ka'bah, which is the Navel of the Earth, whose growth is comparable to a Tree, whose branches are four in number, and are called, the first, 'Sajeratul Mentahar,' and the second 'Taubi,' and the third, 'Khaldi,' and the fourth 'Nasrun Alam,' which extend unto the north, south, east, and west, where they are called the Four Corners of the World. (Walter Skeat, *Malay Magic* 1900)

In Malay, the archaic word *ka'ba* means "cube." Kabah and Ka'aba are not just related to the English words *cube* and *cab* but also to the Egyptian *khabs* (star), *khab* (chariot), the Sumerian brick god Kabta[69] and the Arabic Al Rukaba (North Star). Most importantly, all these are ancient representations of the Hebrew-Aramaic Merkabah, the Throne-Chariot of God. An old meaning of the prefix *mer* is "pure, central, essential."[70]

The Son-Stone

Behold, I lay in Zion a stone for a foundation, a tried stone, a precious cornerstone, a sure foundation; whoever believes will not act hastily.
—Isaiah 28:16 (NKJV)

The Son of God, the Word who became flesh, is the Foundation Stone Himself. He was with His people in the desert. "They drank from the spiritual rock that accompanied them, and that rock was Christ" (1 Corinthians 10:4). Both the words *eben* (stone) and *ben* (son) derive from the verb *banah* (to build), for it is the son that builds the family name. Jesus is the builder of God's family. "Jesus has been found worthy of greater honor than Moses, just as the builder of a house has greater honor than the house itself. For every house is built by someone, but God is the builder of everything" (Hebrews 3:3, 4):

> "You are built upon the foundation of the apostles and prophets with Christ Jesus Himself the chief Cornerstone. In Him the whole structure is joined (bound, welded) together harmoniously, and it continues to rise (grow, increase) into a holy temple in the Lord [a sanctuary dedicated, consecrated, and sacred to the presence of the Lord]." (Ephesians 2:20–21 AMP)

> As you come to him, the living Stone—rejected by humans but chosen by God and precious to him— you also, like living stones, are being built into a spiritual house to be a holy priesthood, offering spiritual sacrifices acceptable to God through Jesus Christ. For in Scripture it says,

> "See, I lay a stone in Zion, a chosen and precious cornerstone, and the one who trusts in him will never be put to shame."

"Now to you who believe, this stone is precious. But to those who do not believe, the stone the builders rejected has become the cornerstone," and, a stone that causes people to stumble and a rock that makes them fall. They stumble because they disobey the message—which is also what they were destined for. (I Peter 2:4–8)

Chosen of God but rejected by His own people, Jesus is "The stone which the builders rejected (that) has become the chief cornerstone. The Lord has done this and it is marvelous in our eyes" (Psalm 118:22–23).

Endnotes

Chapter 1: The Sign of God

[1] Claire O'Kelly, *Concise Guide to New Grange,* 14, 15.

[2] Josephus, *Antiquities of the Jews,* book 1:4.

[3] *Pirke de Rabbi Eliezer,* chapter 3.

[4] Jack Finegan, "Crosses in the Dead Sea Scrolls," *Biblical Archaeology Review* 5, no. 6 (1979): 41, 42, 49.

[5] Genesis Rabba 22, Angelo Rappoport, *Ancient Israel* 197.

[6] Menahoth 74b, Kerioth 5b, Finegan op. cit. 42.

[7] Finegan op. cit. 42.

[8] Ibid., 44, 48, 49.

[9] Epistle to the Ephesians, Chapter 19 *Three Celebrated Mysteries* (trans. Roberts-Donaldson)

[10] Stephen Langdon, "Semitic Mythology" *Mythology of All Races,* Vol. 5. 93

[11] EAW Budge, *The Egyptian Book of the Dead* lxxxiii (83).

[12] Budge, *An Egyptian Hieroglyphic Dictionary,* volume 2: 871, 872.

[13] Ibid. 870.

[14] Ibid. Volume 1: cxliv, 153.

[15] Bundahish 27. 2, 3, Hymn to Tishtrya (trans. E. W. West) *Sacred Books of the East.*

[16] Burr Cartwright Brundage, *The Fifth Sun* 109, 110.

[17] Ibid. 220.

[18] John Bierhorst, *Mythology of South America* 97.

[19] Budge, *An Egyptian Hieroglyphic Dictionary,* volume 2: 823; volume 1: 164, 165,

[20] Ibid. Volume 1: 124, 125.

[21] Ibid. Volume 1: 130.

[22] HB Alexander, "North American Mythology," Vol. 10 *Mythology of All Races* 96.

[23] Ibid. 221, 222, 58.

[24] Ibid. 80.

[25] Caroline Humphrey, and Piers Vitebsky, *Sacred Architecture* 15.

[26] Arthur Cotterell, *MacMillan Illus. Encyclopedia of Myths and Legends* 169

27 Steve Glassman and Armando Anaya, *Cities of the Maya of Seven Epochs* 105.
28 Ibid. 18, 23.
29 Nimuedaju, *Die Sagen von der Religion der Apapocuva-Guarani* 393 in Bierhorst, op. cit. 54.
30 David Talbott, *The Saturn Myth* 144.
31 Dennis Tedlock, *Popul Vuh* 334.
32 Macrobius, *The Saturnalia*, book 1.7.34.
33 Odette Bruhl, "Japanese Mythology," *New Larousse Encyclopedia of Mythology* 417.
34 Floyd Hyatt Ross, *Shinto: The Way of Japan* 45.
35 Joseph Campbell, *Oriental Mythology* 78.
36 James G. Frazer, *The Golden Bough*, volume 1.1:330.
37 Ross, op. cit. 77.
38 Donald MacKenzie, *German Myths and Legends* 3.
39 Gale R. Owen, *Rites and Religions of the Anglo-Saxons* 55.
40 Budge, *An Egyptian Hieroglyphic Dictionary* 350.
41 Warren op. cit. 179.
42 Morris, Jastrow, *Religious Belief in Babylonia and Assyria* 85, 88.
43 Sandars, NK trans. *The Epic of Gilgamesh* 124; Langdon, op. cit. 104.
44 James Legge, *The Tao Te Ching* volume 1. 170.
45 *Bundahish* Chapter 7:14; John R. Hinnells, *Persian Mythology* 23.
46 *The Laws of Manu* 1, 5.
47 *The Mahabharata* (trans. van Buitenen) 462; Yves Bonnefoy, *Mythologies* 875.
48 Micea Eliade, *Images and Symbols* 63.
49 Sheila Moon, *A Magic Dwells* 188.
50 HB Alexander, "North American Mythology" op. cit. 203, 117, 300.
51 Raymond Van Over, *Sun Songs, Creation Myths from around the World* 35.
52 Chapter 37:3, Section 23. 88.
53 Budge, *An Egyptian Hieroglyphic Dictionary*, volume 2: 693.
54 Ross, op. cit. 87.
55 Christie, *Chinese Mythology* 58; Allen, *Star Names, Their Lore and Meaning* 290.
56 L. Wieger, *Chinese Characters* 68.
57 Ibid. 368, 369.
58 John O'Neill, *The Night of the Gods* 657.
59 Wieger, op. cit. 216.
60 Uno Holmberg, *Finno-Ugric Mythology*, 140; *Siberian Mythology* 395, 360.
61 Leinani Melville, *Children of the Rainbow* 33, 110.
62 Ibid. 20,
63 Ibid. 94.
64 Mudrooroo, *Aboriginal Mythology* 52, 110.
65 Mircea Eliade, *Australian Religions* 7, 11, 14.

[66] Ibid. 5–7, 112.
[67] JR Church and Gary Stearman, *The Mystery of the Menorah* 29, 30.
[68] Lenz, Mary Jane, "Star Gods of Ancient America," Archaeology, volume 35 #6.

Chapter 2: The Throne of God

[1] JE Cirlot, A Dictionary of Symbols 40.
[2] EAW Budge, An Egyptian Hieroglyphic Dictionary 530.
[3] Hymns of the Rg Veda, trans. Ralph W. Griffeth 155, 279, 149.
[4] CAS Williams, Outlines of Chinese Symbolism and Art Motifs 425.
[5] Ou-I-Tai, "Chinese Mythology," New Larousse Encyclopedia of Mythology 402.
[6] John O'Neill, The Night of the Gods 586, 587, 605.
[7] Mircea Eliade, Patterns in Comparative Religion 148.
[8] Robert Graves, The White Goddess 292.
[9] Budge, An Egyptian Hieroglyphic Dictionary 686.
[10] Aeschylus, Agamemnon 2. 224, Harvard Classics.
[11] Jellinek ed., Bet ha-Midrasch: Sammlung kleiner Midraschim.
[12] Cited in George Frederick Kunz, The Curious Lore of Precious Stones 236.
[13] The Mahabharata (trans. van JAB Buitenen) 462.
[14] Proincias MacCana, Celtic Mythology 124.
[15] Ibid.
[16] William Warren, Paradise Found 13, 276.
[17] The Voyage of Bran (trans. by Kruno Meyer).
[18] Plato, "Phaedo" 109a-114c The Dialogues of Plato, trans. Jowett
[19] Kunz, op. cit. 237, Graves, Greek Gods and Heroes 29.
[20] Wieger, Chinese Characters 744.
[21] "The Chinese Repository" 7, 519 cited in Warren op. cit. 144.
[22] Anthony Christie, Chinese Mythology 69.
[23] Williams, Outlines of Chinese Symbolism 354, 355.
[24] M. Kerrigan, "Visitors from Heaven" The Diamond Path: Tibetan & Mongolian Myth 88, 91.
[25] Joscelyn Godwin, Arktos, the Polar Myth 96, 97.
[26] Ibid. 95, 100.
[27] Masaharu Anesaki, "Japanese Mythology" Mythology of All Races, Vol. 8, 242.
[28] Stierlin, Architecture of the World-Japan 54.
[29] Budge, Osirus and the Egyptian Resurrection 1. 111.
[30] Ibid. 1. 32, 93, 2. 4, 67.
[31] Ibid. 1. 15; CP Tiele, History of the Egyptian Religion 44.
[32] John Weir Perry, Lord of the Four Quarters 205.
[33] Brundage, The Fifth Sun 196.

34   Kojiki 138.

35   Kerrigan, op. cit. 88.

36   Zev Vilnay, Legends of Jerusalem 15, 157.

37   Rg Veda book 6, Hymn 9

38   Cirlot, op. cit. 379.

39   Van Over, Sun Songs, Creation Myths from Around the World 359.

40   Budge, The Babylonian Legends of Creation, Fourth Tablet lines 43, 44.

41   Roslyn Poignant, Oceanic Mythology 73.

42   Newman, The Hill of the Dragon 74; Alexander, "North American Mythology" 156, 163.

43   Wieger, op. cit. 594.

44   Irene Nicholson op. cit. 69, 70; Brundage, op. cit. 56, 57.

45   Cottie Burland, The Gods of Mexico 176.

46   Nicholson op. cit. 74; James Bailey, The God-kings and the Titans 59.

47   Brundage op. cit. 205.

48   Tedlock, Popol Vuh 52, 300.

49   Ibid. 50, 53.

50   Waters, The Book of the Hopi 118.

51   Max Fauconnet, "Mythology of the Two Americas" New Larousse Enc. of Myth. 443.

52   Harald Osborne, Mythology of South America 84.

53   Ross, op. cit. 185.

54   The Dialogues of Plato, "The Statesman," trans. Benjamin Jowett 23.

55   Edward G. Parrinder, African Mythology 37.

56   Eliade, Australian Religions 1.

57   Richard Heinberg, Memories and Visions of Paradise 106, 107.

58   Ibid. 78.

59   Wieger, op. cit. 647.

60   Bonnefoy, op. cit. 988.

61   Donald Mackenzie, German Myths and Legends 177, 178.

62   Paul Shebesta, Bambuti-Pygmaen vom Ituri, 3, 165.

63   David and Margaret Leeming, A Dictionary of Creation Myths, 291, 292.

64   Ibid. 39.

Chapter 3: The Chosen Place of God

1   Pyramid Texts, Budge, Osirus and the Egyptian Resurrection volume 2. 315.

2   Ross, Shinto: The Way of Japan 17; Anesaki, "Japanese Mythology" Vol. 8 378.

3   Trans. Kisari Mohan Ganguli 396.

4   Berossus, cited in Lenormant, The Beginnings of History 413.

5   Snorri Sturluson, The Prose Edda 37.

6   Budge, An Egyptian Hieroglyphic Dictionary 333.

7   John O'Neill, The Night of the Gods volume 2, 610, 611.

8   Ibid. volume 1, 374.

9   Samuel Noah Kramer, Sumerican Mythology 62, 63.

10   Harold Osborne, Mythology of South America 61–63.

11   Thor Heyerdahl, Easter Island-the Mystery Solved 77.

12   O'Neill, op. cit. 360.

13   Ibid. 1. 143, 144, 379.

14   Nigel Pennick, The Games of the Gods 181.

15   Wieger, Chinese Characters 5.

16   CAS Williams, Outlines of Chinese Symbolism & Art Motives 394.

17   Ibid.

18   Sing Li 114; Chung-Yung, 19, 6; O'Neill, op. cit. 522, 523.

19   John C. Ferguson, "Chinese Mythology" Mythology of All Races, Vol. 8, 28.

20   Wieger, op. cit. 338.

21   Ferguson, op. cit. 52, 59.

22   Legge, op. cit. volume 1, 19, 77.

23   Perry, Lord of the Four Quarters 205.

24   Legge, op. cit. volume 1, 298.

25   Angelo S. Rappoport, Ancient Israel 8, Louis Ginsberg, The Legends of the Jews 1, 8.

26   Budge, The Gods of the Egyptians, volume 1. 467.

27   Ibid. 466.

28   Budge, The Gods of the Egyptians, volume 1. 125.

29   Section 2, 17:6 (trans. Basil Hall Chamberlain).

30   Lewis Spence, The Myths of Mexico and Peru [1913] sacredtexts.com.

31   Roger W. Wescott, Predicting the Past 19.

32   Diogenes, Laertius 2, 9.

33   Ch. 64:1–4.

34   Book 2, Chapter 8 192–193.

35   John R. Hinnells, Persian Mythology 22.

36   Ibid. 30.

37   Budge, An Egyptian Hieroglyphic Dictionary cxxxiii

38   Spence, The Myths of Mexico and Peru [1913]

39   Lenormant, The Beginnings of History 68.

40   Plutarch, Lives 58, 59.

41   WR Lethaby, Architecture, Mysticism and Myth, 237–241.

42   Linda Schele/Mary Ellen Miller, The Blood of Kings, 320, 321.

43   Ibid. 17.

44   Tedlock, op. cit. 335.

45   Burr Cartwright Brundage, The Fifth Sun 39, 40.

46   Lounsbury, "Ancient Writing of Middle America" Senner, The Origins of Writing 211, 213.

47   Joscelyn Godwin, Arctos, the Polar Myth 181, 187.

48   (Hag 12b) Jewishencyclopedia.com "Throne"

49   "Menorah" Jewishencyclopedia.com.

50   Agnes Carr Vaughan, Those Mysterious Etruscans 57.

51   Nigel Pennick, Labyrinths 8.

52   O'Neill, op. cit. 775; Budge, Hieroglyphic Dictionary 362.

53   RT Rundle Clark, Myth and Symbol in Ancient Egypt 58.

54   O'Neill, op. cit. 537; WR Lethaby, Architecture, Mysticism and Myth 73.

55   James Legge, The Chinese Classics, volume 3:34; Warren, op. cit. 216.

56   Cottie Burland, North American Indian Mythology 87.

57   Schele-Miller, op. cit. 277, 269.

58   John O'Neill, The Night of the Gods, Volume I (1893) "The North" 425.

59   Masson-Oursel/Morin, "Mythology of Ancient Persia," Larousse 315.

60   HR Ellis Davidson, Scandinavian Mythology 141; Owen, op. cit. 43

61   Warren op. cit. 187, 213.

62   O'Neill, op. cit. 486.

63   Georges Dumezil, Archaic Roman Religion 642.

64   Servius, On Aeneid 2, 693; ONeill op. cit. 425

65   Holmberg, op. cit. 221, 222.

66   Elwin, The Agaria 96 cited in De Santillana/Von Dechend, op. cit. 219.

67   O'Neil op. cit. 725-729

68   Vaughn, Those Mysterious Etruscans 163.

69   Claiborne, The Roots of English 140.

70   Joseph Naveh, Origins of the Alphabet 69, 70.

71   Church & Stearman, op. cit. 186.

72   Vilnay, Legends of Jerusalem 174.

Chapter 4: The Age of God

1   Budge, An Egyptian Hieroglyphic Dictionary 829.

2   Viaud, "Egyptian Mythology," New Larousse Encyclopedia of Mythology 11.

3   Rundle Clark, Myth and Symbol in Ancient Egypt, 69, 103, 264.

4   Ibid. 264.

5   Ovid, Fasti Book 5, line 625, Book 6, lines 31-32

6   Lenormant, op. cit. 151, 152; Dionysius Halicarnassus, Roman Antiquities I:34.

7   Macrobius, The Saturnalia, book 1. 7. 24, 8.3.

8   Sir James G. Frazer, The Golden Bough, volume 6. 306.

9   Macrobius, op. cit. book 1. 7.26.

10   Frazer, op. cit. volume 6. 314–316.

11 Macrobius, op. cit. book 1. 10.1.
12 Guirand, "Greek Mythology" New Larousse Encyclopedia of Mythology, 93.
13 Strabo cited in Lenormant, The Beginnings of History 531
14 Langdon, "Semitic Mythology" 389; WF Albright, Yahweh and the Gods of Canaan 120
15 Ignatius Donnelly, Atlantis 59.
16 Lucius Accius, Annals; Macrobius op. cit. book 1. 7.37.
17 Masson-Oursel/Morin, "Mythology of Ancient Persia" Larousse op. cit. 318.
18 Zend Avesta, Fargard 2; John R Hinnells, Persian Mythology 34, 113.
19 Yasna 9: 4 and 5, Larousse, op. cit. 310.
20 Giorgio De Santillana/Hertha Von Dechand, Hamlet's Mill 146.
21 Veronica Ions, Indian Mythology 24; Micrea Eliade, Images and Symbols 63.
22 Leeming, op. cit. 202, 204.
23 Waters, The Book of the Hopi 11, 12.
24 Bierhorst, The Mythology of South America 237.
25 Lenormant op. cit. 455, 456; Brundage op. cit. 27, 28.
26 Moodrooru, Aboriginal Mythology 51, 52, 73.
27 Eliade, Australian Religions 1, 3, 43.
28 C. Bishop, The Diamond Path 71.
29 Tablet 29. 16.422.
30 Alexander, "North American Mythology" 126.
31 Bonnefoy, Mythologies 998.
32 Van Over, Sun Songs 371.
33 Marvin W. Meyer, The Ancient Mysteries 170.
34 Mudrooroo, op. cit. 175.
35 Hirata: Sect. 6: 29:20.
36 Alexander, "North American Mythology" 206.
37 Burland, North American Indian Mythology 125.
38 MacCana, op. cit. 126.
39 Irene Nicholson, Mexican and Central American Mythology 24.
40 Vail, op. cit. 72.
41 Ibid. 162
42 12. lines 791, 792
43 Joseph Dillow, The Waters above 190.
44 Warren, op. cit. 281.
45 MacKenzie, German Myths and Legends 187.
46 Eliade, Images and Symbols 66.
47 Waters, op. cit. 6.
48 Heinberg, op. cit. 28, 29.
49 M. Eliade, Patterns in Comparative Religion 256.
50 The Mahabharata (trans. van Buitenen) 447.

51   Immanuel Velikovsky, Earth in Upheaval 237–239, 257; Dillow, op. cit. 174, 175.

52   Dillow op. cit. 176, 177.

53   Roger W. Wescott, Predicting the Past 14.

54   Whitcomb/Morris, The Genesis Flood 371, 372.

55   Brundage, The Fifth Sun 124.

56   Nonnus, Dionysiaca Book 7.

57   Dillow op. cit. 102, 103.

58   Louis L. Ginsberg, The Legends of the Jews 5. 55, 152.

59   Stephanie Dalley, Myths of Mesopotamia "Erra and Ishum" 1, 290.

60   Laws of Manu 1. 68–86.

61   Clark, Indian Legends of the Pacific Northwest 15.

62   Dillow, op. cit. 153–156; HB Alexander, North American Mythology, Vol. 10.

63   James G. Scott, "Indo-Chinese Mythology," Vol. 13 Mythology of All Races, 289.

64   Parrinder, African Mythology 44.

65   Spence, The Myths of Mexico and Peru; Graham Hancock, Fingerprints of the Gods 170.

66   Donald W. Patten and Phillip A. Patten, op. cit. 20, 22, 38.

67   John C, Ferguson, "Chinese Mythology," Vol. 8 Mythology of All Races 28.

68   James Legge, The Texts of Taoism 1.26–28, 89.

69   Talmud Sanhedrin 108a; Ginzberg, op. cit. 1.152, 163.

70   Antiquities, book 1, chapter 3:1.72–74 32.

71   Dimmit/Buitenen, Classical Hindu Mythology 38, 39.

72   Yves Bonnefoy, Mythologies 850.

73   Moon, A Magic Dwells 190.

74   Waters op. cit. 12.

75   Tedlock op. cit. 84.

76   Gallenkamp, op. cit. 145.

77   G. Alexinsky, "Slavonic Mythology" New Larousse Encyclopedia of Mythology, 287.

78   FF Bruce, Genesis ed. The International Bible Commentary 120.

79   The Complete Dead Sea Scrolls 131 trans. Geza Vermes.

80   Tanhuma Buber 23, 24; The Complete Dead Sea Scrolls, trans. Geza Vermes 131.

81   Genesis Rabbah 26:7; Rappoport, op. cit. 1.63, 64.

82   Hesiod, Theogany line 968 (trans. Hugh G. Evelyn-White).

83   Poignant, Oceanic Mythology 42.

84   Moodrooru, op. cit. 136.

85   John Phillip Cohane, The Key 201.

86   Brundage, The Fifth Sun 92, 93.

87   Spence, Myths of Mexico and Peru Chapter 3. "Myths and Legends of the Ancient Mexicans."

88   HB Alexander, "North American Mythology" Mythology of All Races, Vol. 10, 108.

89   Ibid. 211.
90   Ibid. 156, 163.
91   Yves Bonnefoy, Mythologies, 1175.
92   Ella E. Clark, Indian Legends of the Pacific Northwest 63, 64
93   Max Fauconnet, "Mythology of the Two Americas," Larousse, 429.
94   Pedro de Cieza de León: Crónica del Péru I1 Chapter 52 cited in Osborne, op. cit. 80.
95   The Egyptian Book of the Dead Chapter CLXXV.
96   Ella E. Clark, op. cit. 136.
97   Frank Waters, op. cit. 13, 14.
98   Plutarch, Isis and Osirus, volume 5: I sect. 13–15 Loeb Classical Library ed., 1936.
99   Waters, op. cit. 13, 14, 15.
100   Sacred Books of the East, volume 25 81–83.
101   Veronica Ions, Indian Mythology 24.
102   Waters, op. cit. 22.
103   Ions, op. cit. 17, 69.
104   John R Hinnells, Persian Mythology 68, 110.
105   Ions, op. cit. 25.
106   Ibid. 72.
107   Waters, op. cit. 26.

Chapter 5: The Mountain of God

1   Eliade, Images and Symbols 43.
2   Edward Geoffrey Parrinder, African Mythology 110.
3   Eliade, Patterns in Comparative Religion 61.
4   Harold Osborne, South American Mythology 78, 85.
5   Kojiki 456, O'Neill op. cit. 908.
6   Alexander, North American Mythology 92, 207.
7   Burland, North American Indian Mythology 125.
8   H. Luquet "Oceanic Mythology" New Larousse Encyclopedia of Mythology, 465.
9   Max Fauconnet "Mythology of Africa," New Larousse Encyclopedia of Mythology 483.
10   Legge, op. cit. 1. 316.
11   Ferguson, op. cit. 33, Whittaker, op. cit. 35.
12   Eliade, The Sacred and the Profane 145, 146.
13   Richard Hinckley Allen, Star Names: Their Lore and Meaning 451, 452.
14   Alan Friedman, "Science Times," The New York Times, September 12, 1989.
15   Velikovsky, Earth in Upheaval 118, 119; Wescott op. cit. 18, 19, 52.
16   John R Hinnells, Persian Mythology 22.
17   Van Over, Sun Songs, Creation Myths from Around the World 76.

18 Clark, Indian Legends 70, 71.

19 Luquet "Oceanic Mythology" Larousse 453; Poignant, op. cit. 99.

20 Max Fauconnet, "Mythology of the Two Americas" 443; Osborne, op. cit. 105.

21 Osborne, op. cit. 69, 70.

22 Tedlock, op. cit. 37, 360.

23 Zend Avesta Fargard 2, Bundahish Chapter 3.

24 John R. Hinnells, Persian Mythology 30.

25 Eliade, Australian Religions 1.

26 Ibid. 29, Elkin, The Australian Aborigines 224.

27 Velikovsky, Earth in Upheaval 71–73; Whitcomb/Morris, The Genesis Flood 128.

28 Joseph Dillow, The Waters Above 218.

29 JE Cirlot, op. cit. 16.

30 Wieger, op. cit. 45.

31 Raphael Patai, Gates of the Old City 267.

32 Curtin, Creation Myths of Primitive America 18, 19.

33 Starkloff, The People of the Center 107

34 Talbott, The Saturn Myth 203.

35 Eliade, Patterns 100, 102.

36 lines 112–114.

37 lines 104, 105.

38 Black/Green, Gods, Demons and Symbols of Ancient Mesopotamia 72, 76.

39 Eliade, Patterns 79.

40 Francois Lenormant, op. cit. 131, 132; Warren op. cit. 126.

41 Eliade, Patterns 101.

43 Sandars, op. cit. 15, 36, 122.

44 C. Leonard Wooley, The Sumerians 142–144.

45 Hooke, op. cit. 43, 54.

46 O'Neill, op. cit. 400.

47 Cirlot, A Dictionary of Symbols 221.

48 Warren, op. cit. 129, 130.

49 Pennick, The Games of the Gods 147.

50 Williams, op. cit. 53.

51 John Michell and Christine Rhone, Twelve Tribe Nations 47, 48.

52 O'Neill op. cit. 909.

53 Cirlot, op. cit. 186.

54 Williams, op. cit. 376.

55 Burland, The Gods of Mexico 161.

56 Brundage, op. cit. 6, 166, 71.

57 Ibid. 25.

58 Bundahish 5:3, 24:17.

59 2. 174, Pahlavi Texts, Part 1 trans. E.W. West

60   John R Hinnells, op. cit. 34.
61   Rundle Clark, op. cit. 58.
62   Veronica Ions, Egyptian Mythology 22.
63   Eliade, Patterns in Comparative Religion 102.
64   Pyramid Texts (trans. KH Sethe) Rundle Clark, op. cit. 117.
65   Finders Petrie, The Pyramids and Temples of Gizeh 128.
66   Hancock, Fingerprints of the Gods 430, 431, 434.
67   Zajac, The Delicate Balance 141–143, 159, 161.
68   Ibid. 163, 164, 166.
69   Dumezil, op. cit. 64.
70   Burland, The Gods of Mexico 43.
71   Brundage, op. cit. 166.
72   Watkins, The Old Straight Track 75.
73   Ibid. 74.
74   FF Bruce ed. The International Bible Commentary 160; Langdon, "Semitic Mythology" 392.

Chapter 6: The Powerful Names of God

1   Leeming, A Dictionary of Creation Myths 255.
2   Max W. Müller, Egyptian Mythology Vol. 13 Mythology of All Races 209, 312.
3   Hubert Howe Bancroft, Native Races of the Pacific States 546.
4   Leeming op. cit.. 87, 88.
5   Courlander, Tales of Yoruba Gods and Heroes 45.
6   Midrash Tanhuma Buber 1:27a; Genesis Rabba 38:1, 6.
7   Targum Jerushalmi; Pirke de Rabbi Eliezer Chapter 11.
8   Sepher Hajashar; Pirke de Rabbi Eliezer Chapter 24.
9   Midrash Genesis Rabbah 38:1, 6; Tanh. Buber 1:50a-b.
10   Nicholson, Mexican and Central American Mythology 136.
11   Kramer, op. cit. 14.
12   Clark, Indian Legends of the Pacific Northwest 43.
13   Frazer, Folklore in the Old Testament 150.
14   Mythology (1916) Vol. 4 Mythology of All Races (1916)
15   Frazer, Folklore in the Old Testament 150
16   Sanhedrin 109a; Rappoport, op. cit. 238.
17   Josephus, Antiquities 11:4.
18   Jastrow, Aspects of Religious Belief & Practice in Babylonia and Assyria 289.
19   Spence, Arcane Secrets & Ancient Lore of Mexico & Mayan Central America 120.
20   Bonnefoy, Mythologies 984.
21   James G. Scott, "Indo-Chinese Mythology," Vol. 13, Mythology of All Races.
22   Genesis Rabba 38, Rappoport, op. cit. 238.

23   Trans. Nurho de Manhar
24   Owen, Rites and Religions of the Anglo-Saxons 55.
25   Burland, North American Indian Mythology 140.
26   Bierhorst, Mythology of South America 155.
27   Moodrooru, Aboriginal Mythology 184.
28   Budge, An Egyptian Hieroglyphic Dictionary 65, 67.
29   Ibid. 67.
30   Georges Dumezil, Archaic Roman Religion 81.
31   Naveh, The Origins of the Alphabet 68.
32   Dumezil, op. cit. 424, 425.
33   Cohane, The Key 135.
34   Bierhorst, op. cit. 154.
35   Waters, The Book of the Hopi 345.
36   Eliade, Australian Religions 50, 51, 103.
37   Owen, Rites and Rituals of the Anglo-Saxons 11.
38   Sandars, NK trans. The Epic of Gilgamesh 119.
39   Budge, An Egyptian Hieroglyphic Dictionary 59, 122, 554.
40   Cohane, op. cit. 23.
41   Wieger. Chinese Characters 318, 744.
42   Ibid. 661.
43   Ibid. 603.
44   Ibid. 5; Cotterell, op. cit. 104.√
45   Budge, An Egyptian Hieroglyphic Dictionary 23, 24.
47   Martin Brennan, The Hidden Maya 21.
48   Hislop, The Two Babylons 95.
49   Cohane, op. cit. 102.
50   Ibid. 106, 107.
51   Ibid. 241.
52   Ibid. 99.
53   L. Wieger, op. cit. 596.
54   Cohane, op. cit. 241.
55   Dumezil, op. cit. 424.
56   Cohane, op. cit. 76.
57   Colin Renfrew. Archaeology and Language 21, 165.
58   Funk, Word Origins and Their Romantic Stories 269.
59   Roger W. Wescott, Predicting the Past 67, 69.
60   Ibid. 71, 77.

Chapter 7: The River of God

Part 1: The Circle on the Face of the Deep

1 Skeat, Dictionary of English Etymology 328.
2 Cirlot, A Dictionary of Symbols 46, 47.
3 Alexander, "North American Indian Mythology" 275.
4 Budge, An Egyptian Hieroglyphic Dictionary 364.
5 De Santillana/Von Dechend, op. cit. 40.
6 Jeremy and Anthony Green, Gods, Demons and Symbols of Ancient Mesopotamia 130, 131.
7 Rg Veda, book 1, Hymn 173.
8 Tertullian, De Corona Militis 7.ii.85.
9 Kerioth 5b, Jack Finegan, "Crosses in the Dead Sea Scrolls" BAR, volume 5 #6 42.
10 Rappoport. op. cit. 1:113.
11 Alfred Trubner Nutt, Cuchulainn, the Irish Achilles 9.
12 Theogany 776, 242.
13 Odyssey BK 10.
14 John Gray, Near Eastern Mythology 17; Langdon, "Semitic Mythology" 104.
15 Bonnefoy, Mythologies 877; Monier-Williams, Sanskrit Dictionary.
16 O'Neill, The Night of the Gods 808, 1032.
17 Eliade, Patterns 202, 203.
18 O'Neill, op. cit. 1031.
19 Ibid; Theodor H, Gaster, Festivals of the Jewish Year 139.
20 MacKenzie, German Myths and Legends 1.
21 Langdon, "Semitic Mythology" 235, 295, 309.
22 Black-Green, Gods, Demons & Symbols of Ancient Mesopotamia 27.
23 trans. William Muss-Arnolt.
24 Budge, An Egyptian Hieroglyphic Dictionary 349.
25 Ibid. Osirus and the Egyptian Resurrection 2.75.
26 Rundle Clark Myth and Symbol in Ancient Egypt 35.
27 Budge, An Egyptian Hieroglyphic Dictionary 350, 351.
28 Budge, The Gods of the Egyptians 1. 284.
29 Rundle Clark, op. cit. 87.
30 O'Neill op. cit. 780, 781.
31 Budge, The Gods of the Egyptians ll.31.
32 Hesiod, Works and Days 1. 267, 268.
33 Rundle Clark, op. cit. 85, 93.
34 Ibid. 218.
35 Budge, An Egyptian Hieroglyphic Dictionary 192, 193.
36 Rundle Clark op. cit. 223.

37 Budge, The Gods of the Egyptians 1. 343–345.
38 Rundle Clark op. cit. 225, 227.
39 Ibid. 46, 47, 77.
40 Müller, Egyptian Mythology 30.
41 Rundle Clark op. cit. 84, 95.
42 Naveh, Origins of the Alphabet 81.
43 O'Neil op. cit. 67, 537; Kojiki 63:11.
44 Williams, Outlines of Chinese Symbolism and Art Motives 446.
45 Langdon op. cit. 39, 105; Kramer/Maier op. cit. 39, 44.
46 O'Neill op. cit. "The Eye of Heaven" 464, 465.
47 Budge, Egyptian Hieroglyphic Dictionary 409; Monier-Williams, Sanskrit Dictionary.
48 Campbell, Occidental Mythology 167.
49 Eliade, Patterns in Comparative Religion 374.
50 Ibid. 370.
51 Clarke, Mysterious World 86.
52 Eliade, The Sacred and the Profane 49; Perry op. cit. 210.
53 Talbott, The Saturn Myth 126, 128.
54 Jastrow, op. cit. 24; Perry, op. cit. 61.
55 trans. Irving Finkel, British Museum.
56 MacCana, Irish Mythology. 58, 117.
57 Julius Caesar, Gallic Wars Book VI 337.
58 Williams, op. cit. 186; Perry op. cit. 210.
59 Brundage, op. cit. 4.
60 Claiborne op. cit. 189.
61 Tedlock, Popul Vuh 237.
62 Dumezil, op. cit. 130.
63 Tacitus, Annals 2.13. XV. 30.
64 Livy, History of Rome 1.18; Plutarch, Lives, "Romulus" 1.121, Tacitus, Annals 2.13. 15. 30.
65 Cicero, On Divination 2.18, 34; On Laws 2.8.
66 Festus, Auguraculum; Livy, op. cit. book 1.16, 18, book 4.18
67 Sir William Smith's Dictionary of Greek and Roman Antiquities 174–177.
68 Starkloff, The People of the Center 60, 65, 85.
69 Ibid. 100 Black Elk Speaks (1932)
70 Davidson, Scandinavian Mythology 101.

Chapter 7: The River of God

Part 2: The Sevenfold Light

[1]   Rabbi Yaakov Culi, The Torah Anthology 53, 54.
[2]   Halley, Halley's Bible Handbook 627.
[3]   Landon, "Semitic Mythology" 152.
[4]   Rappoport, Ancient Israel 21–23; ONMB 1734.
[5]   Jewishencyclopedia.com
[6]   Lilly S. Eouttenberg & Ruth R. Seldein, The Jewish Wedding Book 90.
[7]   Ibid. 119.
[8]   Ibid. 90, 160.
[9]   Ronald Douglas, Scottish Lore and Folklore 100.
[10]   Abbot, The Keys of Power 395.
[11]   Rappoport op. cit. 3.
[12]   Eliade, Patterns in Comparative Religion 398, 399.
[13]   Gaster op. cit. 112.
[14]   Eliade, Patterns 400–402.
[15]   MacCana, Celtic Mythology 126, 127.
[16]   Mitch and Zhava Glaser, The Fall Feasts of Israel 30–32.
[17]   Ibid. 45, 51.
[18]   Ibid.
[19]   Gaster, op. cit. 221, 226, 227.
[20]   Herodotus, History 1:98.
[21]   O'Neill op. cit. 937.
[22]   Ibid. 944.
[23]   An introduction to the Folklore and Popular Religion of the Malay Peninsula (1900)
[24]   Black/Green op. cit. 136, 137.
[25]   Langdon, op.cit. 156; Gaster op, cit. 114.
[26]   MacCana op. cit. 83.
[27]   Pennick, The Games of the Gods 147.
[28]   Ions, Egyptian Mythology 135.
[29]   Budge, Osirus and the Egyptian Resurrection II.157.
[30]   Plutarch, Lives, volume 5 Isis and Osirus 128 Loeb Classical Library 1936 (trans. Thayer)
[31]   Joscelyn Godwin, The Mystery of the Seven Vowels 22.
[32]   Ions, Indian Mythology 46.
[33]   Warren, op. cit. 257, 258.
[34]   John Abbott, The Keys of Power. 301.
[35]   Monier-Williams, Sanskrit Dictionary.

36   8. 411, 20. 5.

37   Cited in O'Neill op. cit. 974.

38   Dumezil, Archaic Roman Religion 230, 565.

39   Vermaseren, Cybele and Attis 52,53

40   Alexander, "North American Mythology" 19

41   Burland, The Gods of Mexico 83.

42   Waters, The Book of the Hopi 90. 144.

43   Cirlot, op. cit. 53, 54.

44   Alexander, "North American Mythology" 187, 207.

45   Tedlock, op. cit. 49, 360.

46   Cirlot op. cit. 285.

47   Ibid. 173.

48   Roger W. Wescott, Predicting the Past 70.

49   Pennick, Labyrinths 3

50   Kramer/Maier, Myths of Enki, The Crafty God 72, 117.

51   O'Neil op. cit. 665.

52   Cirlot op. cit. 173.

53   Loren McIntyre, "The Lost Empire of the Incas" National Geographic, vol. 144 # 6, 754.

54   Michell, The Dimensions of Paradise 24, 26, 27.

55   Pennick op. cit. 24, 29.

56   Cirlot op. cit. 174, 175.

Chapter 8: The Earth of God

Part 1: The Heaven-Earth

1   Eliade, Patterns in Comparative Religion 243.

2   The Kojiki, (trans. BH Chamberlain)

3   Budge, Osirus 1. 88; Clark, Myth and Symbol in Ancient Egypt 117.

4   Budge, The Gods of the Egyptians I. 509.

5   O'Neill op. cit. 71; Bonnefoy, Mythologies 156, 157.

6   Burland, The Gods of Mexico 77, 99, 131; Brundage op. cit. 226.

7   Bonnefoy op. cit. 996; Cirlot, A Dictionary of Symbols 307, 308.

8   Kramer/Maier, op. cit. 226; Dalley, Myths of Mesopotamia 17.

9   Langdon "Semitic Mythology" 308, 309.

10   Bonnefoy op. cit. 156.

11   Mordiros Ananikian/Alice Werner "Armenian & African Mythology" Mythology of All Races, Vol. 7. 93.

12   Pennick The Games of the Gods 34, 38.

13   Leeming, Creation Myths 263.

14   Josephus, Antiquities of the Jews, book 3, chapter 7 (185) 90.
15   Dumezil, Archaic Roman Religion 652.
16   Bundahish 15. 1–30.
17   Kunz, The Curious Lore of Precious Stones 236.
18   Stierlin ed. Architecture of the World 4, 5.
19   Patil, Cultural History of the Vayu Purana 282.
20   Dumezil, op. cit. 313.
21   Ibid. 315.
22   Black-Green, op. cit. 144, 153.
23   Cirlot op. cit. 81.
24   Budge, An Egyptian Hieroglyphic Dictionary 474, 475.
25   Albert M. Potts, The World's Eye 62.
26   Mudruroo, Aboriginal Mythology 176.
27   Williams, Outlines of Chinese Symbolism and Art Motives 195.
28   James G. Frazer, The Golden Bough VI: 29.
29   Zohar II. 96b.
30   Cirlot op. cit. 248.
31   Aeneid 10.100.
32   Ovid, The Metamorphosis, book 1, ll.255-260.
33   The Mahabharata (trans. van Buitenen) 460.
34   O'Neill, op. cit. 514, 515.
35   Graves, The White Goddess 106, 108.
36   Christie, Chinese Mythology 58; Legge, The Texts of Taoism, Appendix VII 312.
37   Warren op. cit. 15, 16.
38   Wilhelm, The Secret of the Golden Flower, A Chinese Book of Life by Oral Tradition, 22, 61.
39   Ferguson, op. cit. 29, 64.
40   Wieger, Chinese Characters 605.
41   Ellen Chen, The Tao Te Ching 180.
42   Budge, The Egyptian Book of the Dead 50, 289.
43   Ions, Egyptian Mythology 63.
44   Budge, An Egyptian Hieroglyphic Dictionary 815.
45   Ibid. 871, 872.
46   Ibid. 41; Talbott, op. cit. 136.
47   Budge, Book of the Dead, cxxxiii-cxxxvii, chapter 82. 338.
48   Wieger, op. cit. 612.
49   The Dialogues of Plato, "Phaedo" (trans. Benjamin Jowett) 249, 250.
50   Parrinder, African Mythology 48, 49.
51   Bonnefoy, Mythologies 48, 49.
52   Waters, The Book of the Hopi 126–129, 133.
53   George Stuart, "Maya Art Treasures," National Geographic, volume 160 #2 227.

[54] Tedlock op. cit. 330, 362, 363.
[55] Moodrooroo, Aboriginal Mythology 31, 57, 58.
[56] Poignant, Oceanic Mythology 66.
[57] Black-Green, Gods, Demons & Symbols of Ancient Mesopotamia 75, 52.
[58] Dalley op. cit. 329.
[59] Eliade, Patterns 374.
[60] Dumezil op. cit. 579, 662.
[61] Cicero, On the Republic 2.11, On Divination 1.17.
[62] O'Neill op. cit. 433.
[63] Cirlot op. cit. 270.
[64] Humphrey/Vitebsky, Sacred Architecture 12.
[65] Kerrigan, "Visitors from Heaven" The Diamond Path op. cit. 90.
[66] Burland, The Gods of Mexico 138.
[67] Pennick, The Games of the Gods 179.
[68] Wescott, Predicting the Past 74.

Chapter 8: The Earth of God

Part 2: The Emerald City

[1]   Ginsberg, The Legends of the Jews 8.
[2]   BhM 2:32–33; Patai, op. cit. 271.
[3]   Lucian, Vera Historia, book 2.
[4]   Kunz, The Curious Lore of Precious Stones 76, 381.
[5]   Cirlot, A Dictionary of Symbols 132.
[6]   Budge, The Egyptian Book of the Dead 254, 255.
[7]   Budge, The Gods of the Egyptian II. 356.
[8]   Ibid. 1. 509, 511.
[9]   Brugsch, Religion und Mythologie 591 cited in Budge, Osirus I.333.
[10] Kunz op. cit. 226.
[11] Campbell, The Hero With a Thousand Faces 333.
[12] Arthur Cotterell, Myths and Legends 90.
[13] O'Neil, The Night of the Gods 398.
[14] Ibid 96.
[15] Langdon, "Semitic Mythology" 94.
[16] Godwin, op. cit. 169, 170.
[17] Ou-I-Tai, "Chinese Mythology," New Larousse Encyclopedia of Mythology, 381.
[18] John Ferguson. "Chinese Mythology" The Mythology of All Races, Vol. 8, 33, 47.
[19] Wieger, Chinese Characters 658, 659.
[20] Cirlot op. cit. 260, 261; Godwin op. cit. 171.
[21] Williams, op. cit. 72, 234–236.

22 Kunz, op. cit. 86, 87.

23 Kojiki, notes, book 1, chapter 37, 134, volume 14, 16 sacredtexts.com.

24 Bonnefoy, op. cit. 997.

25 Bishop, The Diamond Path 63.

26 Joseph Campbell, The Hero with a Thousand Faces 131.

27 Max Fauconnet, "The Mythology of Two Americas" Larousse op. cit. 441.

28 Tedlock, op. cit. 73.

29 Miguel Leon-Portilla, The Pre-Columbian literatures of Mexico 111, 128.

30 Sahagun, cited in Kunz op. cit. 247, 248.

31 Bonwick, Irish Druids and Old Irish Religioons 238–242.

32 Christie, op. cit. 81.

33 Williams, op. cit. 4.

34 Nigel Pennick, The Games of the Gods 132, 183.

35 Spence, Myths of Peru Chapter 7.

36 MacKenzie, German Myths and Legends 21.

37 John R Hinnells, Persian Mythology 27.

38 Vilnay, Legends of Jerusalem 26.

39 Plato, "Phaedo" trans. Benjamin Jowett 100

40 Rundle Clark op. cit. 228

41 Pyramid Texts 457; W. Max Müller Egyptian Mythology 289.

42 Budge, The Egyptian Book of the Dead 294.

43 Campbell, The Hero With a Thousand Faces 335.

44 Virgil, Georgic 1. 338; Smith's Dictionary of Greek and Roman Antiquities 139.

45 O'Neil op. cit. 806.

46 Leeming op. cit. 307.

47 Kathleen Daly, Norse Mythology A to Z, 43.

48 Tonnelat, "Teutonic Mythology." New Larousse Encyclopedia of Mythology, 276.

49 Leeming, Dictionary of Creation Myths 28; J. Bottero, Religion in Ancient Mesopotamia 147.

50 Brundage, op. cit. 109.

51 Budge, The Gods of the Egyptians, volume 1. 165.

52 Budge, Osirus and the Egyptian Resurrection volume 2. 327.

53 Ibid. 1. 35

54 Budge, An Egyptian Hieroglyphic Dictionary 686, 687.

55 Ibid. 694, 695.

56 Glaser, The Fall Feasts of Israel 157

57 Babylonian Talmud, Tractate Succah 5; "Feast of Tabernacles" JewishEncyclopedia.com..

58 "Hosha'na Rabbah" JewishEncyclopedia.com

59 Glaser op. cit. 175, 176

60 Ibid. 182, 182

[61]   Ibid. 185.

Chapter 9: The Garden of God.

[1]   Warren, Paradise Found 50.
[2]   Babylonian Talmud Berakoth 34b; Seder Gan Eden; BhM 3:1327.
[3]   Eden" JewishEncyclopedia.com.
[4]   Bereshith Rabba xxxiii.
[5]   William F. Albright, Yahweh and the Gods of Canaan 94, 95
[6]   Warren op. cit. 27, 28.
[7]   Sheila Moon, A Magic Dwells 188; O'Bryan, The Dîné: Origin Myths of the Navaho Indians.
[8]   "Eden," Jewishencyclopedia.com.
[9]   Vilnay, Legends of Jerusalem. 17, 28, 29.
[10]   Warren op. cit. 221, Josephus, Antiquities, book 1, 6, 4.
[11]   Vail, The Waters above the Firmament 127.
[12]   Gerald Massey, The Natural Genesis 263.
[13]   Budge, An Egyptian Hieroglyphic Dictionary I. 566-569.
[14]   Ibid.
[15]   Ibid.
[16]   Albright, op. cit. 92
[17]   Aeneid 3. lines 103–105.
[18]   MacKenzie, German Myths and Legends 60, 61.
[19]   John Grant, Viking Mythology 47, 66.
[20]   Davidson, Scandinavian Mythology 91.
[21]   Budge, An Egyptian Hieroglyphic Dictionary 96.
[22]   Ibid. 98, 141.
[23]   Budge, The Gods of the Egyptians II. 75; Talbott op. cit. 124.
[24]   Wieger, Chinese Characters 662.
[25]   Perry, Lord of the Four Quarters 211–213.
[26]   Williams, Outlines of Chinese Symbolism 72–26; O'Neil, op. cit. 161.
[27]   The Kojiki, part 1 chapter 37; Nihongi, book 1.
[28]   Juliet Piggott, Japanese Mythology 17; Anesaki, "Japanese Mythology", Vol. 8 212.
[29]   Clio Whittaker, Oriental Mythology 96.
[30]   Ibid. 99, 102.
[31]   Frazer, The Golden Bough (abridged) 27.
[32]   Bierhorst, Mythology of South America 31.
[33]   Scott, "Indo Chinese Mythology," Vol. 12 Mythology of All Races 289.
[34]   Melville, Children of the Rainbow 83.
[35]   Ixtilxochitl, Don Ferdinand d'Alva (c. 1568–1648) "Historia Chichimeca" 1658.
[36]   Leeming op. cit. 80.

37   The Mahabharata (trans. van Buitenen) 58.
38   Hesiod, Works and Days 90; Bonnefoy, Mythologies 423.
39   Waters, The Book of the Hopi 234.
40   HB Alexander, "North American Mythology" 34.
41   Leeming op. cit. 241, 242.
42   Bierhorst op. cit. 66.
43   Ibid. 133.
44   Osborne, op. cit. 38; Bailey, The God-kings and the Titans 83.
45   Immanuel Velikovsky, Earth in Upheaval 83.
46   Brundage, The Fifth Sun 46, 47.
47   Ibid. 166, 171–173.
48   Leeming op. cit. 8, 9.
49   Fauconnnet, "Mythology of the Two Americas" Larousse op. cit. 431.
50   Van Over, Sun Songs 86.
51   Alexander, "North American Mythology" 124.
52   Fauconnet, op. cit. 430.
53   Ibid. 432.
54   McClintock, The Old North Trail 495, 496.
55   Bierhorst op. cit. 66, 68.
56   Bonnefoy, Mythologies 257, 259.
57   Ibid. 258, 259.
58   Ibid.
59   Parrinder, African Mythology 19, 20.
60   Parrinder, op. cit. 38, 42.
61   Heinberg, Memories and Visions of Paradise 34, 53, 54.
62   Parrinder, op. cit. 44.
63   Heinberg, op, cit. 83.
64   Max Fauconnet, "Mythology of Africa" New Larousse Encyclopedia of Mythology, 483.
65   Poignant, Oceanic Mythology 42–44.
66   Ibid. 54.
67   Melville, op. cit. 6, 7.
68   Heinberg, op. cit. 78.
69   John Phillip Cohane, The Key 73,125
70   David and Margaret Leeming, A Dictionary of Creation Myths, 88, 89.
71   Frazer, Folklore in the Old Testament 5,6
72   Graham Hancock, Fingerprints of the Gods 135.
73   Eliade, Patterns in Comparative Religion 383.
74   Ginsberg, op. cit. 1.82.

Chapter 10: The Island of God

[1] John O'Neil op. cit. 31, 32.

[2] Mircea Eliade, Patterns in Comparative Religion 289; AH Sayce, "The Academy" 263.

[3] DM Field, Greek and Roman Mythology 62.

[4] Thor Heyerdahl, Easter Island: The Mystery Solved 110, 111.

[5] CAS Williams, Outlines of Chinese Mythology 233.

[6] Anthony Christie, Chinese Mythology 69.

[7] Walter Skeat, Malay Magic 86, 87, 92.

[8] Ibid. 14.

[9] Mudrooroo, Aboriginal Mythology 20, 21.

[10] Mircea Eliade, Australian Religions 29.

[11] James G Frazer, Folklore in the Old Testament 92, 93.

[12] EA Wallis Budge, The Babylonian Story of the Deluge.

[13] The Mahabharata trans. JAB van Buitenen 75

[14] Morris Jastrow, Religion in Babylonia and Assyria 215.

[15] Withnell, Customs and Traditions of the Aboriginal Natives of Northwestern Australia.

[16] Frazer, The Golden Bough 173.

[17] Robert Graves, The White Goddess 109.

[18] Homer, Odyssey trans. At Murray 1. 14 ff.

[19] Budge, The Gods of the Egyptians I.107; Book of the Dead 254.

[20] O'Neil, op. cit. 803, 804.

[21] Whittaker, Oriental Mythology 102; Kojiki SECT. XC, Part XV.

[22] Lynd, History of the Dakotas cited in Donnelly, Atlantis 96.

[23] Bundahish, chapter 10.

[24] Leinani Melville, op. cit. 48, 91, 102.

[25] Lynd op. cit. 96.

[26] Sam Gill, Native American Religions 12, 13.

[27] Clark, Indian Legends From the Pacific Northwest 141.

[28] Nicholson op. cit. 131.

[29] Brinton, The Annals of the Cakchiquels 14, 25–26.

[30] HB Alexander, Latin American Mythology 181, 182.

[31] Anesaki, "Japanese Mythology," Mythology of All Races Vol. 8, 222, 223.

[32] Ibid. 378.

[33] Ignatius Donnelly, Atlantis 177.

[34] MacKenzie, German Myths and Legends 11, 12.

[35] Donnelly op. cit. 271, 324.

[36] Warren op. cit. 184.

[37] Homer, Odyssey, book 9.528.

38   Settegast, Plato Prehistorian, 91.
39   De Santillana/Von Dechend, op. cit. 363.
40   O'Neill op. cit. 31, 32.
41   Herodotus, History, book iv 184.
42   O'Neill op. cit. 33.
43   Vishnu Purana, book 2, chapter 2, 166, 167 trans. Horace Hayman Wilson sacred-texts.com.
44   Ibid.
45   Bonwick, Irish Druids and Old Irish Religions 300.
46   Ibid. 293.
47   Donnelly, op. cit. 24.

Chapter 11: The Wings of God

1   Zev Vilnay, Legends of Jerusalem 172
2   The Zohar 2:8a, Raphael Patai, Gates to the Old City 432, 433
3   JE Cirlot, A Dictionary of Symbols 27, 28
4   L. Wieger, Chinese Characters 37
5   Mircea Eliade, Patterns in Comparative Religion 41, 42
6   John O'Neil, The Night of the Gods 751, 763
7   Ibid. 758
8   Williams, Outlines of Chinese Symbolism and Art Motives 320
9   Dennis Tedlock trans, The Popul Vuh, 336
10   Cirlot, op. cit. 253, 254
11   Roger W. Wescott Predicting the Past 40
12   Williams, op, cit, 323, 325
13   John C. Ferguson, "Chinese Mythology" 99
14   John R Hinnells, Persian Mythology 22. 40
15   Masson-Oursel & Morin 'Mythology of Ancient Persia" Larousse 315
16   Volume 1, Part 4
17   EAW Budge, The Gods of the Egyptians Vol. 2. 96
18   Wescott op. cit. 26
19   Rundle Clark, op. cit. 37, 245
20   Budge, The Gods of the Egyptians Vol. 1. 466
21   Ibid. Vol. 2. 97
22   HB Alexander, "North American Mythology" Mythology of All Races, Vol. 10. 59
23   Leeming, A Dictionary of Creation Myths 245
24   Alexander, op. cit. 24
25   Max Fauconnet, New Larousse Encyclopedia of Mythology 427
26   Clark, Indian Legends of the Pacific Northwest, 162
27   Nicholson, Mexican and Central American Mythology 78,79

28 Burr Carwright Brundage, The Fifth Sun 104, 106
29 Ibid. 109
30 Ibid. 105,123
31 Burland, The Gods of Mexico 161,162
32 Ibid. 150,153
33 Brundage op. cit. 78,142
34 Harold Osborne, South American Mythology 137, 140
35 HB Alexander, "Latin American Mythology" 194
36 John Bierhorst, Mythology of South America 58
37 Raymond Van Over, Sun Songs 51
38 Ibid. 88, 91
39 Katherine Berry Judson, Myths and Legends of Alaska 39, 59
40 Corcoran, "Celtic Mythology" Larousse 231
41 Nigel Pennick, The Games of the Gods 237
42 Moodruroo, Aboriginal Mythology 23,24
43 Cicero de Divinatione 34; de Legibus 8
44 Plutarch, Lives, Romulus 161
45 Cicero de Div. 1.2, II.17, 26, 34; Smith's Dictionary of Greek & Roman Antiquities 175, 176
46 Pennick, op. cit. 19
47 O'Neill op. cit. 434, 435, 440
48 Black/Green, Gods, Demons and Symbols of Ancient Mesopotamia 107
49 Stephanie Dalley, Myths of Mesopotamia 33
50 Ibid. 222, 318
51 The Mahabharata trans. JAB van Buitenen 88
52 Dennis Tedlock The Popul Vuh 36, 37
53 Ibid. 89, 90, 91
54 Ibid. 92, 93, 94, 360
55 Alexander, Latin American Mythology 174, 177
56 Rundle Clark, op. cit. 217
57 Alexander, North American Mythology 65
58 Osborne, op. cit. 126
59 Bierhorst, The Mythology of South America 165
60 Mudrooroo, Aboriginal Mythology 75
61 Holmberg, "Siberian Mythology" 357, 447

Chapter 12: The Rock of God

1   B. Talmud Yoma 54b.
2   Sode Raza, Yalkut Reubeni on Genesis 1, 4. See also Ginzberg I.12.
3   O'Neill, The Night of the Gods 129.

4   Cirlot, A Dictionary of Symbols 314; Eliade, Patterns 226, 227.

5   Ginzberg, Legends of the Jews, volume 4, 96.

6   Eliade, Patterns in Comparative Religion 219, 220.

7   Ibid. 221, 222, 223.

8   John R Hinnells, Persian Mythology 89.

9   O'Neill op. cit. 121.

10   Eliade, Patterns 235.

11   Suidas s.v. Agyiai (trans. Suda Byzantine Greek lexicon tenth century AD).

12   Pausanias, Description of Greece 1. 44. 2.

13   Graves, The White Goddess 355.

14   Eliade, Patterns 235.

15   Bonnefoy op. cit. 897.

16   G. Alinsky "Slavonic Mythology" New Larousse Encyclopedia of Mythology, 293.

17   Macrobius, The Saturnalia book 1. 19. 14, 15.

18   Griffith, Rg Veda 113.

19   Pythian Ode 6.1.1.3, 4.

20   Graves, op. cit. 377.

21   Walter F. Otto, Dionysus Myth and Cult 157.

22   Hesiod, Theogony lines 463, 494, 625; O'Neill op. cit. II. 825.

23   Hesiod, Theogony lines 493–500.

24   G. Dumezil, Archaic Roman Religion 176.

25   Festus, In Voce, Lapidem Silicem; O'Neill op. cit. 112.

26   Dumezil op. cit. 178.

27   F Guirand-AV Pierre "Roman Mythology" New Larousse Encyclopedia of Mythology 213.

28   Dumezil op. cit. 366.

29   Rundle Clark op. cit. 39.

30   Ions, Egyptian Mythology. 36.

31   Ibid. 44.

32   Dalley, op. cit. 129.

33   Kunz, op. cit. 230, 231.

34   Bonwick, op. cit. 161.

35   Ibid. 313–320.

36   Ibid. 223.

37   MaCana op. cit. 117.

38   Bonwick op. cit. 36, 212.

39   Eliade, Patterns 233.

40   Michell/Rhone, Twelve Tribe Nations 160.

41   Watkins, The Old Straight Track 31.

42   Eliade, Patterns 225.

43   MacKenzie op. cit. 60, 61.

44   Olav Tausch (Own work) [GFDL CC BY 3.0

45   Abbott, The Keys of Power: A Study of Indian Ritual and Belief 240, 242.

46   Ibid. 243.

47   Ibid. 241.

48   Ibid. 244.

49   Wescott, Predicting the Past 22.

50   Ions, Indian Mythology 13, 42.

51   O'Neill op. cit. 686.

52   Brundage, The Fifth Sun 8–12, 224.

53   Harold Osborne South American Mythology 69, 70.

54   Ibid. 61, 81.

55   McIntyre, "The Lost Empire of the Incas" 737, 746.

56   Brinton, Annals of the Cakchiquels 43, 47.

57   Piggott, Japanese Mythology 48, 49.

58   O'Neil op. cit. 192.

59   Stierlin, Architecture of the World-Japan 83.

60   Poignant, op. cit. 72.

61   Ibid. 100.

62   Luquet, "Oceanic Mythology" New Larousse Encyclopedia of Mythology, 457.

63   Joscelyn Godwin, Arktos, the Polar Myth 102.

64   Eliade, Images and Symbols 40.

65   O'Neil op. cit. 117.

66   Hastings, James, ed. Encyclopedia of Religion and Ethics 665.

67   Hinnells, op. cit. 51.

68   Stephen Langdon, "Semitic Mythology" Mythology of All Races, Vol. 5, 16.

69   Samuel Noah Kramer, Sumerian Mythology 61

70   O'Neill op. cit. 55

Bibliography

Holy Bibles

- The Holy Bible, New International Version ® (NIV), copyright © 1973, 1978, 1984, 2011 by Biblica, Inc.™ Used by permission. All rights reserved.
- The New King James Version (NKJV), copyright © 1982 by Thomas Nelson, Inc. Used by permission. All rights reserved.
- The Amplified® Bible (AMP), copyright © 1954, 1958, 1962, 1964, 1965, 1987 The Lockman Foundation Used by permission." (www. Lockman.org).
- New Revised Standard Version Bible (NRSV), copyright © 1989 the Division of Christian Education of the National Council of the Churches of Christ in the United States of America. Used by permission. All rights reserved.
- The King James Bible (KJV), public domain.
- Douay Rheims 1899 American Edition (DRA), public domain.
- The One New Man Bible, Revealing Jewish Roots and Power (ONMB), © 2001 True Potential Publishing. Used by permission. All rights reserved worldwide.
- Orthodox Jewish Bible (OJB), copyright © 2002, 2003, 2008, 2010, 2011 by Artists for Israel International.

Books

Abbott, John. *The Keys of Power, A Study of Indian Ritual and Belief.* Secaucus, New Jersey: University Books 1974.

Albright, WF. *Yahweh and the Gods of Canaan.* University of London, 1968.

Alexander, HB. "Latin American Mythology." Volume 11 of *Mythology of All Races*.

New York: Cooper Square 1916.

———. "North American Mythology." Vol. 10 of *Mythology of All Races* New York: Cooper Square: Cooper Square 1916.

Allen, Richard Hinckley. *Star Names, Their Lore and Meaning*. New York: Dover 1963.

Anesaki, Masaharu "Japanese Mythology" *Mythology of All Races*, Vol. 8 New York: Cooper Square 1928

Angus, S. *The Mystery-Religions*. New York: Dover Publications, 1975.

Astrov, Margot. *American Indian Prose and Poetry*. Boston: Beacon Press, 1992.

Bailey, James, *The God-kings and the Titans*, New York: St. Martin's Press 1973.

Bancroft, Hubert Howe, *Native Races of the Pacific States* New York: D. Appleton 1874.

Barton, George A, *Archaeology and the Bible*, Revised Philadelphia: 1925.

Bierhorst, John, *Mythology of South America*, New York: William Morrow & Co. 1988.

Black Elk, Wallace and William S. Lyon, *Black Elk, The Sacred Ways of a Lakota*. San Francisco: Harper & Row, 1990.

Black, Jeremy and Anthony Green, *Gods, Demons and Symbols of Ancient Mesopotamia* Austin: University of Texas Press 1992.

Bonnefoy, Yves, *Mythologies*, volumes I and II Chicago: University of Chicago Press 1991.

Bonwick, James *Irish Druids and Old Irish Religions* 1894 Dorset Press USA: 1986

Bottero Jean, *Religion in Ancient Mesopotamia*, Chicago: University of Chicago 2001.

Brennan, Martin, *The Hidden Maya* New Mexico: Bear and Co. 1998.

Brinton, Daniel G. *The Annals of the Cakchiquels*, Library of Aboriginal America vi Philadelphia: 1885.

Bruce, FF, Genesis Ed. *The International Bible Commentary* Grand Rapids: 1986.

Brugsch, Heinrich, *Religion and Mythology of the Egyptians* Leipzig, 1887

Brundage, Burr Cartwright, *The Fifth Sun* Austin, Austin: University of Texas Press 1979.

Budge, EAW *Osiris and the Egyptian Resurrection* Volumes 1 and 2 London: 1911 reprinted New York: Dover 1973.

————*The Egyptian Book of the Dead* London 1895 reprinted New York: Dover 1967.

————*An Egyptian Hieroglyphic Dictionary,* London 1920, New York Dover 1978.

————*The Gods of the Egyptians* London 1904 reprinted New York: Dover 1969.

————*The Babylonian Legends of Creation* London 1921 sacred-texts.com.

Bühler, George trans, The Laws of Manu Hong Kong: Forgotten Books, 2008.

Bullfinch, Thomas, *Mythology,* New York: Crown Pubishers Inc. 1979

Burland, Cottie A. *North American Indian Mythology,* New York: Peter Bedrick 1985.

————*The Gods of Mexico,* New York: GP Putnam's Sons 1967.

Burnet, Thomas, *Sacred Theory of the Earth* 1691 sacred-texts.com.

Caesar, Julius, *The Gallic War,* trans. HJ Edwards, Loeb Classical Library, Cambridge Massachusetts: Harvard University Press 1917.

Campbell, Joseph, *The Hero with a Thousand Faces* Princeton NJ: Prinston University Press 2004.

————*The Masks of God: Occidental Mythology* New York: Viking Press 1964.

————*Oriental Mythology* London: Souvenir Press, 1962.

Clement of Alexandria, *The Exhortation to the Greeks*, trans. GW Butterworth. Loeb Classical Library, Cambridge Massachusetts Harvard University Press 1919.

Chen, Ellen M., *The Tao Te Ching: A New Translation and Commentary.*New York: Paragon House 1989.

Christie, Anthony, *Chinese Mythology,* New York: Peter Bedrick Books 1985.

Church, JR and Gary Stearman, *The Mystery of the Menorah* Oklahoma City: Prophecy Publications 1993.

Cicero, *The Nature of the Gods*, trans. PG Walsh, New York: Oxford University Press 1997.

Cirlot, JE, *A Dictionary of Symbols,* London: Philosophical Library 1962.

Claiborne, Robert, *The Roots of English*, New York: Times Books 1989.

Clark, Ella E. *Indian Legends of the Pacific Northwest* Berkeley: 1953.

Clark, RT Rundle, *Myth and Symbol in Ancient Egypt,* London: Thames & Hudson 1959.

Coe, Michael D., *Breaking the Maya Code,* New York: Thames and Hudson 1993.

Cohane, John Phillip, *The Key,* New York: Crown Publishers Inc. 1969.

Cotterell, Arthur, *MacMillan Illustrated Encyclopedia of Myths and Legends,* New York: MacMillan Co. 1989.

Courlander, Harold, *Tales of Yoruba Gods and Heroes,* New York: Crown Publishers 1973.

Culi, Rabbi Yaakov, *The Torah Anthology,* trans. by Rabbi Aryeh Kaplan, New York: Mosnaim Publishing Corp.1977.

Curtin, Jeremiah, *Creation Myths of Primitive America* 1898, etext at sacred-texts.com

Dalley, Stephanie, *Myths of Mesopotamia: The Creation, The Flood, Gilgamesh,* New York: Oxford University Press 1991.

Daly, Kathleen N., *Norse Mythology A to Z,* New York: Facts on File Inc. 1991.

Dante, Aligheri, *The Divine Comedy,* trans. Charles Eliot Norton London and New York: 1902.

De Buck, Adriaan, *The Egyptian Coffin Texts* Chicago: Oriental Institute Pub. 1939

De Manhar, Nurho, *The Sepher Ha-Zohar* New York: 1900–1914 sacred-texts.com.

Dillow, Joseph C., *The Waters Above* Chicago: Moody Press 1981.

Dimmitt, Cornelia and Van Buitenen, JAB. *Classical Hindu Mythology: A Reader in the Sanskrit Puranas,* Philadelphia: Temple University Press 1978.

Diodorus Siculus, *Library of World History,* trans. CH Oldfather, Cambridge, Massachusetts Loeb Classical Library, Harvard University Press 1989.

Donnelly, Ignatius, *Atlantis,* New York: Harper Brothers 1949

Douglas, MacDonald Ronald, *Scottish Lore and Folklore* New York: Beekman House 1982.

Dumezil, Georges, *Archaic Roman Religion,* Volumes. 1,2, Chicago: University of Chicago Press 1966.

Eliade, Mircea *Images and Symbols* Princeton: Princeton University Press 1969.

———*Australian Religions,* Ithaca New York: Cornell University Press 1973.

———*Patterns in Comparative Religion,* New York: Penguin Books 1958.

———*The Sacred and the Profane,* New York Harcourt Brace & World 1959.

Elworthy, Frederick Thomas, *The Evil Eye* New York: Bell Publishing Co 1989.

Field, DM, *Greek and Roman Mythology* New York: Chartwell Books 1977.

Ferguson, John C. "Chinese Mythology" *Mythology of All Races,* Vol. 8 New York: Cooper Square 1928

Frazer, Sir James G., The Golden Bough, Vols. I-13 London: MacMillan 1906–1915.

———Folklore in the Old Testament, London: MacMillan 1923

Gantz, Jeffrey, trans. The Mabinogion New York: 1976.

Gaster, Theodor H, Festivals of the Jewish Year New York: Morrow Quill Paperbacks,1978.

———The Dead Sea Scriptures in English Translation, Garden City NY: Doubleday 1964.

Gifford and Sibbick, Warriors, Gods and Spirits from South American Mythology New York: Peter Bedrick Books 1983.

Gill, Sam D., Native American Religion, 2nd Ed.,Canada: Wadsworth Thompson 2005.

Ginsberg, Louis L. The Legends of the Jews Volumes. I-VI Philadelphia: 1925. sacred texts.com

———Legends of the Bible, New York: Jewish Publication Society 1956.

Glaser, Mitch and Zhava, The Fall Feasts of Israel, Chicago: Moody Press 1887.

Glassman, Steve and Armando Anaya, Cities of the Maya in Seven .. Jefferson: North Carolina, McFarland and Co 2011.

Godwin, Joscelyn, Arktos, the Polar Myth Grand Rapids, MI: Phanes Press, 1993.

———The Mystery of the Seven Vowells Grand Rapids, MI: Phanes Press 1991.

Grant, John, Viking Mythology, New York: Chartwell Books 1990.

Graves, Robert The White Goddess New York: Creative Age Press 1948.

———Greek Gods and Heroes New York: Laurel-Leaf Books 1977.

Gray, John, Near Eastern Mythology, New York: Peter Bedrick Books 1969

Griffith, Ralph TH, trans. Hymns of the Rg Veda New Delhi, India: Motilal Banarsidass Publishers Pvt. Ltd. (1999).

Halley, Henry H. Halley's Bible Handbook, Grand Rapids: Zondervan 1961.

Hancock, Graham, Fingerprints of the Gods, New York: Crown Publications 1995.

Hastings, James, ed. An Encyclopedia of Religion and Ethics, 13 Volumes, New York: Scribner and Sons 1908–1926.

Hays, HR, In the Beginnings, New York: Putnam 1963.

Heinberg, Richard, Memories and Visions of Paradise, Los Angeles: Jeremy P. Tarcher Inc. 1989.

Herodotus, History of Herodotus, Vol. 6, Great Books of the Western World, Edited Robert Maymard Hutchins, Encyclopedia Britannica Chicago 1952.

Hesiod, *The Works and Days, Theogony, The Shield of Herakles*, trans. Richmond Lattimore, Ann Arbor: University of Michigan Press 1991.

Hinnells, John R., *Persian Mythology*, New York: Peter Bedrick Books 1985.

Hislop, Alexander, The Two Babylons Neptune New Jersey: Loiseaux Brothers 1959

Homer, *The Iliad*, trans. AT Murray Loeb Classical Library Cambridge, Massachusetts, Harvard University Press New York: 1919.

———*The Odyssey*, trans. SH Butcher and A. Lang New York: 1888.

Hooke, SH, *Babylonian and Assyrian Religion* University of Oklahoma Press 1963.

Heyerdahl, Thor, *Easter Island: The Mystery Solved*, New York: Random House 1989.

Humphrey, Caroline and Piers Vitebsky, *Sacred Architecture*, New York: Barnes & Noble Books 2005.

Ions, Veronica, *Indian Mythology*, New York: Peter Bedrick Books 1983.

Jastrow, Morris, *Aspects of Religious Belief & Practice in Babylonia and Assyria* New York: Barnes and Noble 1971.

Jones, Prudence/Nigel Pennick, *A History of Pagan Europe* New York: Routledge1995.

Jones, Sir William, *Asiatic Researches,* volume 2 London 1801.

Reprint: Kessinger Publications 2007,

Josephus, *The Works of Josephus*, Peabody, Massachusetts: Hendrickson Publishers 1987.

Judson, K. Berry, *Myths and Legends of Alaska* Chicago: AC McClurg & Co. 1911.

———*Myths & Legends of California & the Old Southwest* Chicago: AC McClurg 1912.

Kato, Genchi, *A Study of Shinto*, New York: Barnes & Noble 1971.

Kerrigan, M, Bishop, C, Chambers J, *The Diamond Path: Tibetan & Mongolian Myth* Amsterdam: Time-Life Books 1988.

Kramer, Samuel, Noah *Sumerian Mythology* revised edition Philadelphia: 1961.

Kramer SN and John Maier, *Myths of Enki, the Crafty God*, Oxford University Press New York: 1989.

Kuno, Meyer, trans. *The Voyage of Bran,* London: Alfred Nutt 1895.

Kunz, George Frederick, *The Curious Lore of Precious Stones* New York: Bell Publishing Co. 1989.

Langdon, Stephen, *Semitic Mythology*, New York: Cooper Square 1931.

Leeming, David and Margaret, *A Dictionary of Creation Myths,* New York:

Oxford University Press 1994.

Legge, James, trans. *The Texts of Taoism,* New York: Dover Publications 1962.

Lenormant, Francois, *The Beginnings of History,* Paris: Charles Scribner's Sons 1882.

Leon-Portilla, Miguel, *Pre-Columbian Literatures of Mexico,* Norman, Oklahoma: University of Oklahoma 1969.

Lethaby, WR, *Architecture, Mysticism, and Myth,* New York: MacMillan and Co. 1892.

Livy, *History of Rome* Books 1–5, trans. Valerie M. Warrior Indianapolis: Hackett 2006.

MacCana, Proinsias, *Celtic Mythology,* London: Hamlyn Books 1970.

MacKenzie, Donald A., *German Myths and Legends,* Cambridge, Massachusetts, Crown Publications 1985.

Macrobius, *The Saturnalia,* trans. Robert A. Kaster, Cambridge, Massachusetts, Harvard University Press 2011.

Massey, Gerald, *The Natural Genesis,* London: Williams and Norgate 1883.

Matt, Daniel, Zohar, *the Book of Enlightenment,* New York: Paulist Press 1983.

Melville, Leinani, *Children of the Rainbow* Wheaton, Illinois: Theosophical Publishing House 1969.

Meyer, Marvin W, *The Ancient Mysteries,* New York: Harper & Row 1987.

Michell, John/Christine Rhone, *Twelve Tribe Nations,* Grand Rapids: Phanes Press 1991.

Michell, John, *The Dimensions of Paradise* Rochester, Vermont 1988.

Moodrooru, *Aboriginal Mythology,* Aquarian-Harper Collins London 1994.

Moon, Sheila, *A Magic Dwells, A Poetic & Psychological Study of the Navajo Emergence Myth* San Francisco: Guild for Psychological Studies Pub. House, 1985.

Morris, Henry M./John C. Whitcomb, *The Genesis Flood* Phillipsburg, New Jersey: 1961.

Müller, Max W. *"Egyptian Mythology" Mythology of All Races,* Vol. 13 New York: Cooper Square 1918.

———Editor, *Sacred Books of the East,* Oxford University Press 1879–1910
sacred-texts.com, The Internet Sacred Text Archive.

Naveh, Joseph, *Origins of the Alphabet,* London: Cassell 1975.

New Larousse Encyclopedia of Mythology, Richard Aldington, Delano Ames, trans. Robert Graves; Félix Guirand; National Art Library, Archive of Art and Design. London: Chancellor Press, 1996.

Nicholson, Irene, *Mexican and Central American* Mythology New York: Peter Bedrick Books 1987

Nutt, Alfred Trubner, *Cuchulain, the Irish Achilles* London: D. Nutt 1900.

Nuttall, Zelia, *The Codex Nuttall,* New York: Dover Publications 1975.

O'Kelly, Claire, *Concise Guide to New Grange* Cork, Ireland: 1991.

O'Neill, John, *The Night of the Gods,* Volumes I and II, London: Harrison & Sons 1893.

Osborne, Harold, *South American Mythology,* New York: Peter Bedrick Books 1968.

Otto, Walter F., *Dionysus Myth and Cult,* Bloomington: Indiana University Press 1965.

Ovid (Publius Ovidius, Naso) *The Metamorphoses,* trans. FJ Miller, London: 1960.

Loeb Classical Library Cambridge, Massachusetts, Harvard University Press

————*Fasti,* trans Betty Rose Nagle Bloomington: Indiana University Press 1995.

Owen, Gale R., *Rites and Religions of the Anglo-Saxons* Dorset Press 1985.

Parrinder, Edward Geoffrey, *African Mythology* New York: Peter Bedrick 1982.

Patai, Raphael, *Gates to the Old City* Northvale, NJ: J. Aronson: 1988.

Patten, Donald Wesley, *The Biblical Flood and the Ice Epoch,* Seattle: Pacific Meridian Publishing Co. 1966.

Pausanias, *Description of Greece,* trans. WHS Jones, Internet Ancient History Sourcebook.

Pennick, Nigel, *The Games of the Gods* York Beach, Maine, Rider and Co.1989.

————*Labyrinths* Runestaff - Old England 1986.

Perry, John Weir, *Lord of the Four Quarters,* New York: MacMillan Co. 1966.

Petrie, WM, Flinders, *The Pyramids and Temples of Giseh* London: 1883.

Philippi, Donald L., trans. *The Kojiki* Tokyo: Tokyo University Press 1979.

Piggott, Juliet, *Japanese Mythology,* New York: Peter Bedrick 1969.

Pindar, *Odes,* trans. Roy A. Swanson New York: Ardent Media 1974.

Plato, *The Dialogues of Plato,* trans. Benjamin Jowett, Oxford: Clarendon Press 1953.

Plato in Twelve Volumes, trans. Harold North Fowler. Cambridge, Massachusetts: Harvard University Press, William Heinemann Ltd. London 1921.

Platt, Rutherford H. Jr., editor, *The Forgotten Books of Eden* Cleveland, Ohio: 1981.

Plutarch, *The Lives of the Noble Grecians and Romans,* volume XIV, Great Books of the Western World, ed. Robert Maynard Hutchins,

Pliny the Elder, *Natural History,* London: Penguin Classics 1991.

Poignant, Roslyn, *Oceanic Mythology* Books Feltham, Middlesex: Newnes Books, 1985

Potts, Albert M., *The World's Eye*, University Press of Kentucky 1982.

Rappoport, Angelo S., Ancient Israel, Myths and Legends Vols. I-III New York: 1987.

Renfrew, Colin, Archaeology and Language New York: Cambridge University Press 1987.

Ross, Floyd Hiatt, *Shinto: The Way of Japan* Boston: Beacon Press 1965.

Routtenberg, Lily S. and Ruth R. Seldin, *The Jewish Wedding Book* New York: 1967.

Sandars, NK, trans. *The Epic of Gilgamesh* London: Penguin Books 1964.

Schele-Miller, *The Blood of Kings*, London: Sotheby's, Fort Worth in ass with Kimbell Art Museum 1986.

Scott, *Indo Chinese Mythology, "Mythology of All Races"* Volume 13 New York: Cooper Square 1916.

Senner, Wayne N., ed., *The Origins of Writing* Lincoln: University of Nebraska Press 1991.

Settegast, Mary, *Plato Prehistorian,* Cambridge, Massachusetts: Lindisfarne Press 1987.

Skeat, Walter, *The Concise Dictionary of English Etymology,* Hertfordshire, UK: Wordsworth Editions 1993.

————*Malay Magic: An Introduction to the Folklore and Popular Religion of the Malay Peninsula,* New York: Macmillan and Co. 1900.

Smith, William LL.D, *A Dictionary of Greek and Roman Antiquities,* London: John Murray 1875.

Spence, Lewis, *Arcane Secrets & Ancient Lore of Mexico & Mayan Central America* Detroit: B. Ethridge-Books 1973.

Starkloff, Karl, *The People of the Center,* New York: The Seabury Press 1974.

Sturluson, Snorri, *The Prose Edda,* trans. Jean I Young Berkeley: 1954.

Tacitus, Censorinus, *De Die Natale,* trans. William Maude, New York: The Cambridge Encyclopedia Co. 1900.

————*The Annals and the Histories,* volume XV, Great Books of the Western World, Ed. Robert M. Hutchins, Encyc. Britannica (Chicago 1952).

Talbott, David N., The Saturn Myth, New York: Doubleday 1980.

Tedlock, Dennis, trans. *The Popol Vuh* New York: Simon and Schuster 1985.

Tiele, Cornelis Petrus, *History of the Egyptian Religion* trans. James Ballingall, London: 1982.

Vail, Isaac Newton, *The Waters above the Firmament* Cleveland: Clark & Zangerle 1886,

Van Buitenen Johannes Adrianus Bernardus trans, *The Mahabharata* Chicago: University of Chicago Press 1973.

Van Over, Raymond, *Sun Songs, Creation Myths from Around the World* New York:

New American Library 1980.

Vaughn, Agnes Carr, *Those Mysterious Etruscans,* New York: Doubleday 1964.

Velikovsky, Immanuel, *Earth in Upheaval,* New York: Delta 1965.

Vermaseren, Maarten J. *Cybele and Attis, Myth and Cult* London: Thames and Hudson 1977.

Vermes, Geza, *The Complete Dead Sea Scrolls in English* revised edition, London: Penguin Books 2004.

Vilnay, Zev, *Legends of Jerusalem,* Philadelphia: Jewish Publication Society 1973.

Vine, WE, *Vine's Complete Expository Dictionary,* Nashville: Thomas Nelson 1996.

Virgil, *Eclogues, Georgics, Aeneid, Great Books of the Western World* Chicago: 1952.

Ward, William Hayes, *The Seal Cylinders of Western Asia* Washington D.C.: Carnegie Institution of Washington 1919.

Warren, William F, *Paradise Found,* Cambridge, Massachusetts: Riverside Press 1885.

Waters, Frank, *Book of the Hopi,* Penguin Books New York: 1977.

Watkins, Alfred, *The Old Straight Track* London: Abacus 1925.

Wescott, Roger Williams, *Predicting the Past: An Explanation of Myth, Science and Prehistory,* Deerfield Beach, Florida: Kronos Press 2000.

West, EW, trans. *Bundahish, Sacred Books of the East,* London: Oxford University Press 1897.

Whitcomb, John C. and Henry M. Morris, *The Genesis Flood* Phillipsburg, New Jersey Presbyterian and Reformed Publishing Co. 1961.

Whittaker, Clio, ed., *Oriental Mythology,* Secaucus, New Jersey: Chartwell Books 198).

Wieger, L., *Chinese Characters,* New York: Dover Books 1965.

Wilhelm, Richard, *The Secret of the Golden Flower: A Chinese Book of Life by Oral Tradition*, trans. Cary F. Baynes; London: Kegan Paul, Trench & Trubner 1931.

Williams, CAS, *Outlines of Chinese Symbolism and Art, Motives* New York: Dover 1976.

Wright, David trans. *Beowulf,* New York NY: Penguin Classics 1957.

Periodicals

De Vaux, Roland, "Was There an Israelite Amphictony?" *Biblical Archaeology Review*, volume III #2 1977.

Finegan, Jack, "Crosses in the Dead Sea Scrolls," *Biblical Archaeology Review* volume V #6 1979.

Ilan, Zvi and Damati, Emmanuel, "The Synagogue at Meroth," *Biblical Archaeology Review, volume* XV #2 1989.

Lenz, Mary Jane, "Star Gods of Ancient America," *Archaeology*, volume 35 #6.

Lemaire, André, "Who or What Was Yahweh's Asherah?" *Biblical Archaeology Review*, volume X #6 1984.

Mazar, Eilat, "Royal Gateway to Ancient Jerusalem Uncovered," *Biblical Archaeology Review*, volume XV #3 1989.

McIntyre, Loren, "The Lost Empire of the Incas," *National Geographic* Volume 144 #6 December 1973.

McDowell, Bart, "The Aztecs," *National Geographic*, volume 158 #6 December 1980.

Muss-Arnolt, "The Urim and Thummim," *American Journal of Semitic Languages*, July 1900, 199–204.

Severy, Merle, "The Celts," *National Geographic*, volume 151#5 May 1977.

Stager, Lawrence E., "The Shechem Temple," *Biblical Archaeology Review*, July/August, volume 29 #4.

Stuart, George E., "Maya Art Treasures Discovered in Cave," *National Geographic* Volume 160 #2 August 1981.